Claudia Hammond has presented two series of the critically acclaimed *Emotional Rollercoaster* on BBC Radio 4, as well as numerous other series on science and medicine. She reports regularly for Radio 4's *Woman's Hour*, has degrees in Applied Psychology and Health Psychology and lectures for British and American universities. This is her first book.

From the reviews of *Emotional Rollercoaster*:

'[A] lovely book ... There is much more than just science here. Hammond leavens her account of the latest scientific research with plenty of other material drawn from art, philosophy and her own everyday experiences ... [A] wealth of fascinating observations ... Refreshing ... Humour, sensitivity and warmth emanate from every page' *Guardian*

'Hammond summarises a formidable array of current research, weaving this gently between personal anecdotes, detailed accounts of some of the more lurid and entertaining experiments and nuggets of hard fact' *Sunday Telegraph*

'Claudia Hammond takes us on a neuroscientific tour ... Leaving the mushy metaphysical stuff to the poets, the book treats emotions as rational, material processes ... Hammond's style is accessible and anecdotal, and her refusal to romanticise emotions is bracing' *Financial Times*

'An entertaining, informed guide to the responses that drive and colour our lives' *Independent*

'Recommended . . . Hammond's conversational style is lively and entertaining, and her eagerness to ground theory in real-life examples hooks the reader from the outset . . . An entertaining yet comprehensive overview of emotions research . . . *Emotional Rollercoaster* is a most enjoyable journey . . . bound to appeal to a wide audience' *The Psychologist*

CLAUDIA HAMMOND

Emotional Rollercoaster

A Journey Through the Science of Feelings

HARPER PERENNIAL
London, New York, Toronto and Sydney

Harper Perennial
An imprint of HarperCollins*Publishers*
77–85 Fulham Palace Road
Hammersmith
London W6 8JB

www.harperperennial.co.uk

This edition published by Harper Perennial 2006

First published in Great Britain by Fourth Estate 2005

A catalogue record for this book is available from the British Library

ISBN-13 978-0-00-716467-7

p. 87, Bill Viola, *The Quintet of the Astonished*, 2000. Video/Sound Installation. Colour video rear projection on screen mounted on wall. Photo: Kira Perov
pp. 97, 120, 145, 148, 172, 201, 316, 330 © Getty Images
p. 121 Courtesy of Mütter Museum, College of Physicians of Philadelphia
p. 254 © Corbis
p. 295 © Miles Barton/Naturepl.com
For additional information on illustrative content see p. x

Extract from *Peribanez* by Lope de Vega, trans. Tanya Ronder
Extract from Primo Levi reprinted with permission of Simon & Schuster Adult Publishing Group from *The Drowned and the Saved*. Translated from the Italian by Raymond Rosenthal. English translation copyright © 1988 by Simon & Schuster, Inc.

Typeset in Minion with DIN display by
Rowland Phototypesetting Ltd, Bury St Edmunds, Suffolk

For Nick and Bonnie

contents

acknowledgements

So many people have been extraordinarily generous with their time in answering my questions and advising me on how to go about writing this book. I've also been touched by the number of people who have sent me journal articles or lent me books.

I've long been interested in this subject and originally suggested the idea of a series on the science of emotions to Jane Ellison, commissioning editor at BBC Radio 4. I'm very grateful to her for commissioning a first and then a second series on the subject which I was lucky enough to make with the endlessly patient and brilliant producer Marya Burgess.

For advice and encouragement on finding a publisher I must thank my extraordinary agent David Miller, as well as Chris Paling, Anna McNamee and Annalisa Barbieri.

The following people generously gave up their time to discuss their work with me, some of them long before I began writing this book, but their ideas added hugely to my understanding of the subject: Randy Cornelius, Dylan Evans, Mark McDermott, Mary Douglas, Hannah Steinberg,

Christina Jones, Joanna Hawthorne, Chris Peterson, Nicola Green, Val Curtis, Mary Phillips, Joseph LeDoux, Paul Stenner, Riccardo Draghi-Lorenz, Elaine Hatfield, Nakia Gordon, Andreas Bartels, William Falls, Paul Ekman, Paul Rozin, Ernie Noble, Anne Manyande, Andy Field, Rob Briner, Janet Polivy, Alyson Bond, Phil Cowen, Simon Gelsthorpe, Robert Frank, Karl Grammar and William Morgan. Many of these people also lent me books or sent me useful articles and references, as did John Oates and John Aggleton.

There is also a long list of people who were kind enough to share their emotional experiences with me (some have chosen to be referred to by their first names in this book): Chloe, Maggie and David, Gillian, Karl, Grant Dalbeth, Christine Stewart, Andrew, Ben Gillow, Alex, Rachel Hedley, Jo Morris, Sandy Raffan, Sally Mackay, Nigel Jackson, James Marshall, Eric Moody, Steph West, Alan Murray, and Reg, Marjory and Julia Chapman who were particularly open and honest regarding their emotional experiences.

Jo Morris, Jo Saxby, Matt Taylor, Nicola Green and Alan Finch have all lent me photographs and particular thanks go to Sandy Raffan and Adam Tjolle for taking photos specially for the book.

I'm very grateful to Georgina Laycock and Philip Gwyn Jones at Flamingo for commissioning the book and to publisher, Nicholas Pearson, and editor, Nick Davies, at 4th Estate for their excellent suggestions. I'd particularly like to thank editor, Kate Hyde, for tactfully making such incisive improvements to the book. Thank you to Samantha Noonan for indexing so enthusiastically.

Friends and family also helped in all sorts of ways. My

mother Bonnie was particularly encouraging and my father Nick Hammond answered my natural history queries. Antonia and David Sturt-Hammond, Ursula Saunders, Dylan Evans and my partner Tim read early drafts of various chapters and I thank them for all their useful suggestions.

introduction

germany, october 1944

Reg lay down on his bunk in the German prisoner of war camp to read the longed-for letter which had finally arrived in the latest Red Cross parcel. In the officers' camp at Eichstatt near Munich he and the other 600 prisoners slept twenty to a room. Much of the time he was both cold and hungry. It was the letters from Marjory, the stunning dancer he had met at the outbreak of the war, that kept him going. But for three months he had heard nothing.

The dapper, dark-haired army officer had first spotted her when she was one of eight dancing girls in a Max Miller show at the Holborn Empire in London. She looked like a film star with her mass of blonde hair and her long, long legs. Had he not needed to catch the last train home straight after the show, he would have waited outside the theatre in the hope of meeting her. She was gorgeous and he was besotted, but would she even have noticed him among the other stage door Johnnies? There must be some way he could get to know the lovely dancer.

* * *

He was in luck. A neighbour of his parents knew Marjory and promised to pass on a message. Might he be permitted to write to her? Yes, she would like that. And perhaps they could meet for afternoon tea? Yes, that would be very nice. Reg was overjoyed, sensing that Marjory was the girl for him. But the war was to intervene. Before that first date his territorial army regiment was suddenly dispatched to Lille in France so, determined to remain optimistic, the young subaltern arranged to meet her when he came home on leave two months later. The day of their date soon arrived but Reg did not. Marjory heard that Reg was listed as missing in action and at first feared that he had been killed. However, soon the news came through that he had been captured on 12th June 1940 at St-Valery-en-Caux, just south of Dieppe. Now he was a prisoner of war in Germany. Theirs was to be a love affair by letter.

Usually Marjory's letters lifted Reg's spirits but as he opened the envelope he sensed something was wrong. There was something not quite right about the handwriting; it wasn't Marjory's, it was her mother's, and she was writing with bad news. With the London shows long since closed and determined to contribute to the war effort, Marjory had been working as a convoy officer with the Auxiliary Territorial Service. She lived with twenty-five other young women in a manor house in Kent and each day they set off to collect military trucks, ensured they were filled with the correct equipment and drove them off in a long line, always observing the mandatory distance between each truck, to wherever the army required them. But one morning Marjory had such bad stomach pains that she was unable to work so the other girls left without her. A couple of hours later she heard the familiar pulsating motor of a doodlebug overhead. It sounded like a small plane chugging along through the sky until the engine suddenly cut out. She waited to see where it would drop, praying that it wasn't on her. She was used to that slow ten or eleven seconds where you were temporarily suspended between life and the possibility of imminent death until you heard the explosion some distance away and for you, at least, life continued. But this time things were different. The bomb fell directly onto the roof of the manor house, collapsing all three storeys deep down into the cellars. Marjory was left trapped under the rubble, pinned down by an oak beam which had crushed her legs. As she lay in agony amid the debris of the building she tried to scream but couldn't. She listened in silent terror as the neighbours explained to the firemen that luckily the house was empty because all the girls were out at work. No, they couldn't leave her here. Don't go! As she heard the firemen's voices become more distant, she managed to cry out in panic, but even after her screams were heard, each brick and beam had to be lifted off by hand and it was another four hours before they reached her. Realising she was pinioned by the beam with both her back and pelvis broken the emergency services decided there was only one way to get her out – and Marjory overheard their plans: 'There's nothing for it, we'll have to amputate her legs.' She refused to let them, demanding they find another way. She didn't care how much it hurt. They did manage to extricate her but when she was finally examined in hospital the news was bad; as well as five fractures to her pelvis, her back was broken in two places resulting in paralysis from the waist down.

Reading the letter in his bunk in the prisoner of war camp Reg imagined poor Marjory trapped beneath the rubble and was reminded of the fear of his own experience of being buried alive only months before. He had been in a group digging an escape tunnel when the officer in front of him accidentally hit a live cable. The roof of the tunnel collapsed on top of them, leaving him lying face down, trapped in the earth, stones and dust that had showered on top of them. He was terrified and called out to the man in front. Then he heard voices behind him. His friends were crawling in through the debris to rescue them. He felt them grab his ankles and then the pain as they dragged him out backwards, the skin ripping off the front of his shins and knees. When he eventually reached daylight he realised that apart from his grazed legs he was uninjured, but when they hauled his friend out next, his body was limp; he had been electrocuted.

Putting the frightening memories out of his mind, Reg concentrated on what had happened to Marjory. Anger replaced his fear. The Germans could do whatever they wanted to him, but how could they drop bombs on a lovely girl like Marjory? How could they do this – leaving her paralysed when she'd done nothing wrong? He left his bunk, rushed out of the room to find some German guards and then spewed abuse at them. As a result he found himself locked in an underground cell in solitary confinement. Usually this punishment didn't bother him. In a crowded camp with 600 officers time to himself was rare and so a spell in the underground cell held a certain peace not available elsewhere in the camp, but this time he spent the lonely hours alternating between fury with the Germans and sadness that he couldn't even comfort poor Marjory. There was a third emotion too and that was hope. He wished more than he had ever wished for anything that she might get better.

While Marjory was in hospital, equally hopeful and determined that she would learn to walk again, Reg remained in Oflag VI. B, waiting for the war to end. During the day the other ranks left their camps to go out to work on local farms, but the Geneva Convention forbade Reg and the rest of the officers from working, with the exception of 'voluntary work' – collecting fir cones to put on the fire. The remainder of the time their main distraction came from planning the next escape attempt. After years of malnutrition Reg knew it wouldn't be him escaping; you needed to be strong to stand a chance, but even helping others to escape gave him hope. Despite the years of boredom and deprivation, he and his fellow POWs even experienced moments of joy. 'What kept us going was seeing the funny side of things. Such little things really. If a German's hat blew off on the parade ground, we all clapped and laughed.' They particularly looked forward to the 'special items' which were smuggled into the camp. Among these were fountain pens and shaving brushes, or at least that's what they appeared to be. In fact they contained a chemical with an indelible, noxious smell. The POWs would blunder into passing German guards, press the pen against them and activate the plunger, thus saturating their uniforms with the stinking chemical. With only one uniform apiece and no way of removing the chemical, the guards were forced to go about their duties smelling foul – a blow for German pride and a boost for British morale.

Emotions such as joy and sadness would come and go, but for Reg two others – anger and hope – remained constant. Anger, at the German military for what they had done to Marjory and for their treatment of his fellow POWs; hope – unshakeable and steadfast – that the Allies would win the war and that he would soon be home. Indeed, amongst the British POWs hope was virtually compulsory. If anyone did show signs of despair the other prisoners would soon cheer them up. 'Chin up' was the rule. Winning the war was surely just a matter of time. And there were reasons to be cheerful. One day soon, he knew that he and Marjory – the beautiful dancer he had loved for years but had never properly met – would have that first date.

The tale of Reg and Marjory is a love story, but it's also a tale of anger, disgust, joy, fear, sadness and hope. Emotions are central to their story. They made the experience what it was.

During much of the twentieth century, while other areas of psychology expanded, the emotions were not considered important topics for psychological research. They tended to be thought of as rather undefined concepts that had a deleterious influence on our behaviour by disrupting rational thinking. Emotions lacked appeal as a subject for study because they were considered difficult to measure and quantify; instead time was spent studying more rigorously-defined topics like memory, perception and learning. An explosion in research on emotions in the last ten years has brought a dramatic change to the field.

However, the idea still persists that life would be better, if only we weren't at the mercy of our feelings, that they are ready to emerge at any moment, forcing us to say things we don't mean and to desire things we would be better off without. Once emotions are examined more closely, with the aid of research from neuroscientific, psychological and biological perspectives, it becomes clear that each emotion is both wiser and more useful than we might expect.

In this book I've chosen nine emotions to explore in detail. There is disagreement amongst emotions researchers as to which feelings can be classified as true emotions. This selection covers the nine emotions which I believe are key – the emotions which have the most to tell us both about the way we deal with the world and how our brains interact with the rest of our bodies. At first sight it appears that more of these emotions are negative than positive, but on this journey through the emotions it soon becomes clear that feelings cannot be categorised as simply good or bad.

For the most part I am concentrating on the everyday experience of emotions, rather than the extremes. It's clear that research has plenty to teach us about our own feelings. Just as the body responds to strong emotion with a racing heart or butterflies in the stomach, so our physiology also feeds into our emotions. For example, laboratory experiments have demonstrated that forming your facial muscles into a smile can make you feel happier, even if you're unaware that you are smiling. With advances in brain scanning, and crucially for researchers, the reduced costs of using scanners, it is now possible to observe the way the brain responds when a particular emotion is experienced. It is not the case that one part of the brain lights up for fear and another for anger, but each emotion employs different combinations of brain systems. Such investigations are beginning to shed light on the purpose of each emotion, but this progress in the understanding of the relationship between the brain's chemicals and the way we feel does not mean that we are doomed to be ruled by our emotional brains, programmed with immutable responses to every situation. Events in the outside world and even the way we choose to think about those events can still influence those chemical responses. Only by considering emotions at all these levels can we begin to see the whole picture.

One extraordinary aspect of emotions is that, despite massively differing circumstances, the way an emotion feels remains constant across the lifespan. While an older man's joyful moment incorporates a lifetime of complex memories and thoughts, the feeling is the same for a delighted toddler – and both the toddler and the old man share the facial expression which communicates that joy.

Not all emotions are detectable nor probably even present from birth. Just as a baby develops in other areas of life, so their emotional understanding changes as they grow. I've traced the emotions from basic emotions such as joy and disgust to those we develop as we grow older, such as guilt, and finally an emotion I believe to be an essential part of life – hope. On the way I'll examine such varied topics as the purpose of tears, why the length of your earlobes might be related to jealousy, why people are loath to drink water containing their own spit and whether hope can make you live longer.

one

joy

Less than a week after newspapers claimed that as a result of global warming snow might never be seen in London again, I woke one Wednesday in January 2003 and everything sounded slightly muffled – no pushchair wheels catching on the bumpy paving stones, no metal trolleys jangling against the kerb outside the shop. It could only mean one thing – snow. The only sounds which were louder than usual were the cries of children, suddenly inspired to skip to school, thrilled with the free entertainment. Working at home that day, I was distracted by watching snowflakes from my desk and wondering whether the snowflakes were fatter than they used to be. By the afternoon I could resist the snow no longer and my partner and I walked up to Primrose Hill, a small green park from which you can see the pointed cages of London Zoo and beyond it, London's landmarks laid out in a line – the Dome, Canary Wharf, the London Eye and the Post Office Tower. We'd assumed that as it was a Wednesday afternoon the park would be fairly empty, but in fact there were hundreds of people. Reaching the flat circle of tarmac at the top of the hill, we detected something unusual about the atmosphere, different even from sunny weekends when people bring kites and picnics. That difference was the presence of joy – pure joy at the unexpected snowfall. Strangers were grinning at each other with the kind of festive, communal happiness you might expect to find on New Year's Eve. People climbed the hill carrying cardboard boxes and plastic bags – anything, so long as they could slide. Despite the absence of children there was already a coterie of snowmen, one with tens of sticks for spiky hair. Three men had rather optimistically brought skis and were side-stepping their way up the hill, accompanied by a local newspaper photographer hoping to catch them in action. A couple were tearing off bin bags from a roll and handing them out to anyone brave enough to throw themselves down the hill. You could see from their smiles that everyone was thrilled. This was unexpected, snatched joy. And these weren't children. The children were all in school, but while they worked, adults were out playing. A group started rolling a massive snowball. People joined in and moments later a snowball of *James and the Giant Peach* proportions was rolled off the side of the hill, soon smashing against a tree on its way down and disintegrating, to a delighted cheer from the crowd.

This joy was unplanned but provided we have the opportunity we all invest time and effort into the creation of joyful moments – saving up for holidays, arranging to see friends, finding somewhere nice to eat – with the hope of creating happy times. Food, sex or even a hot bath can make you feel happy, but there seems to be more to joy than the physical pleasure. Joy is all-encompassing, whether it lasts for a moment or for longer. One little boy I interviewed told me that after he had won a football match he had felt joyful for exactly three days.

If people are asked to list emotions, happiness is often the first one they name. Most emotions researchers include it in their list of basic emotions, but they tend to refer to it as joy because joy has more of a moment-to-moment quality. While happiness can be a background emotion, joy indicates the quality of feeling at one particular moment.

the importance of new sandals

Corblets beach was always our favourite – a large sweep of sand overlooked by a crenellated fort at one end of the Channel island of Alderney. When the tide was out we would lie on some sloping rocks, conveniently slanted at just the right angle for reading and sunbathing. When the tide came in, we would retreat to one of the grassy alcoves in the ferns leading up the hill to the fort (or we did until the day a rat emerged from the undergrowth and disappeared with my entire cheese and pickle baguette in its jaws). One July day in 1979, long before we knew about the presence of rats, I left my brand new navy blue sandals on a boulder just below the slanting rocks and went into the sea to argue with my sister. Later, forgetting all about my sandals with the special patterns punched in the straps, we spent the afternoon marking out a badminton court in the sand, playing frisbee and building runnels and dams in the sand. Slowly the tide came in flooding the dams faster than we could build them, but when it was time to leave the beach I was horrified as I saw the boulder had been submerged and imagined my new sandals washing out to sea. This story might have been placed in the chapter on sadness and would have been, had an elderly lady not come up to me, asking if I was looking for something. From one hand she dangled the sandals by their straps. I was delighted they'd been found, but that was nothing compared with my joy at discovering that the elderly lady was Elizabeth Beresford, author of the Wombles books. We were thrilled when she gave us a lift back to the guesthouse in her little white Mini.

What might have been 'the day my sandals were swept out to sea' in family folklore became 'the day Elizabeth Beresford rescued my sandals'. It was that transformation from misery to joy which had made me feel so happy; the possibility of sadness had made the joy stand out.

Back in the sixteenth century the French essayist Michel de Montaigne spotted the immense joy that the absence of pain can bring. On the subject of his agonising kidney stones he wrote, 'Is there anything so delightful as that sudden revolution when I pass from the extreme pain of voiding my stone, and recover, in a flash, the beauteous light of health, full and free, as happens when our colic paroxysms are at their sharpest and most sudden? Is there anything in that suffered pain that can outweigh the joy of so prompt a recovery? Oh how much more beautiful health looks to me after illness.'

When researchers ask people to recount the last time they experienced joy, relief from suffering doesn't tend to be mentioned. Instead the occasions tend to involve seeing friends, eating, drinking, having sex or achieving success. It seems that we don't always notice the joyful moments we experience. The renowned Canadian psychologist Keith Oatley has experimented with different methods of establishing the frequency with which we experience each emotion. When he asked people to keep a diary and note down each time they noticed a particular emotion, he found that negative emotions predominated. If, however, he prompted people at random with a pager, asking them to note down the emotion they were experiencing at that moment, happy times suddenly became twice as frequent as fear or anger. It seems that we are more likely to notice our experience of negative emotion. Moreover the cessation of those negative feelings can lead to an increase in joy.

Researching joy is harder than it sounds. One researcher, Jonathan Freedman, found that people became distinctly uncomfortable when asked to discuss what made them happy. His research assistant found it easier to get people to talk about their sex lives than about joy, as though it were too personal a subject to discuss with a stranger. This reluctance to discuss joy is reflected in the paucity of language surrounding the subject. While there are plenty of words for feeling miserable – wretched, sad, downcast, depressed, dejected, unhappy etc. – there are far fewer words for joy. It simply isn't discussed with the same frequency as negative emotions.

Within literature misery features far more often than joy. As the French novelist Henri de Montherlant said, happiness 'writes white'. The same has happened within psychology. Although we all seek joy and many would say it makes life worthwhile, it doesn't receive the same amount of attention from researchers as negative emotions like fear or sadness, presumably because they cause more problems. The eminent psychologist Martin Seligman has drawn attention to the positive emotions such as joy by pioneering a branch of psychology known as positive psychology. In order to help the subject along, a $100,000 prize is awarded each year for the best research on well-being or on helping people to flourish. Joy is slowly beginning to be seen not as emotion that it would be nice to experience, but one which is essential.

* * *

When I was walking home late at night recently I witnessed some more joy. A man with a German accent was walking along on the other side of the road talking loudly, but ecstatically into his mobile. 'I had to call you. I can't believe how happy I am. These fantastic guys were over from the US office, so we went to a really expensive restaurant in Piccadilly for lunch. Six of us, all on expenses. Such great guys. Here I am in my new job getting taken to fantastic London restaurants. We had all this food and wine and I'm being paid for it. This must be the best job in the world. I'm so excited. I'm so happy I can hardly speak.'

He bounced as he walked. The joy appeared to activate his whole body. It's known that joy does cause the heart rate to rise. It also energises us, compelling us to stand up when we hear good news, in contrast to sitting down when it's bad. With his particular interest in the science of emotions Charles Darwin noted a case of 'psychical intoxication' recorded in the *Medical Mirror* from 1865. A young man opening a telegram containing the news that he had inherited a fortune became initially pale, then exhilarated and restless. He staggered around the streets whilst 'uproariously laughing', talking irritably and singing loudly. Everyone thought he was drunk although he hadn't drunk anything (apparently confirmed after he vomited and the contents were examined). He was simply full of joy.

the joyful brain

In his ecstatic state the man who received the telegram would not have been the ideal person to carry out a task involving careful analysis and precision because our mood can change the way we think. When we feel any emotion strongly enough, whether it's fear or ecstasy, we tend to make decisions without processing information as deeply. Instead we rely on heuristics or stereotypes in our judgements.

A standard psychological test involves giving people a paragraph describing another person. After a delay they are asked questions about the person. If they are experiencing a strong emotion at the time they will rely on stereotypes triggered by certain words in the passage rather than on the actual information. This does have advantages in terms of efficiency which might be essential in a time of crisis. It might be unfair to assume that a man wearing torn clothes walking behind you down a dark alleyway is dangerous, but if that brief moment of anxiety causes you to quicken your pace so that you reach the main road faster, then even if he turns out to be harmless you haven't lost anything by speeding up. Were you to slow down while you considered whether there was in fact anything suspicious about the man's behaviour or whether you were unfairly generalising on the basis of his clothing, you might put yourself in danger. Nevertheless it is not the case that strong emotions always lead to a decrease in accurate reasoning. Emotions tend to be seen as irrational urges which interfere with effective decision-making. Research has shown, however, that in certain circumstances, feeling good can change your thinking in such a way that you make *better* decisions, particularly if those decisions necessitate creativity.

You are given a standard white candle, a box of drawing pins and a book of matches. Your task is to attach the candle to the wall and light it in such a way that the wax won't drip onto the floor. When people were given this task in a laboratory *after* their mood had been manipulated the results were striking. One group had been given a comedy to watch before the task while the others saw a more serious film. Three quarters of the group who had seen the happy film found a workable solution, compared with a measly 13% of those who had seen the serious film. The solution to the task, incidentally, is to use some drawing pins to fix the drawing pin box to the wall as a ledge for the candle. Then you light the candle with the matches and the box successfully catches the drips.

Dr Alice Isen, a psychologist at Cornell University, has conducted years of experiments in this area, finding that mild happiness can improve performance in everyone from children to doctors. She finds cunning ways of making one group of people feel slightly happy – giving them money, telling them they've done unusually well on a task or simply providing fruit juice and biscuits. Compared with another group who hadn't had any reward, the happy group were significantly better at thinking up unusual word associations. In another experiment doctors were given a bag of sweets to cheer them up before they made a diagnosis and extraordinarily this resulted in a better diagnosis where they took more information into account. This had nothing to do with them putting in extra effort because they were happy; instead there appeared to be something qualitatively different in their thinking. When people feel happy they seem to consider problems from more angles.

Joy allows people to think optimistically and to remember other times when they were successful, as well as allowing them to focus on the task at hand. Your negotiation skills even improve when you feel happy, so the time to ask for a pay-rise is not when you're feeling fed up with a job, but when you're feeling good.

If we turn to the chemistry of what's happening in the brain there might be an explanation for these processes. Dopamine is just one of hundreds of neurotransmitters or chemical messengers in the brain, but we know more about it than any of the others. The cells that produce dopamine are only found in a few areas of the brain, mainly in the brain stem, but it has effects in many other areas. Food, sex and drugs all release dopamine, producing feelings of pleasure, but only for a limited time. Like all neurotransmitters dopamine is a chemical that allows one neuron or nerve cell in the brain to communicate with the next and in this way messages are passed between the millions of neurons in the brain. Dopamine is released when we feel joyful and it is these dopamine levels that might affect the way our thoughts are processed. Alice Isen suggests that the release of dopamine into a part of the brain called the anterior cingulate region might help us to switch perspective. Seeing things from another viewpoint is a key component of creativity and we might actually find this easier if dopamine is helping us along while we explore ideas. The same could apply to negotiation; to bargain successfully it's necessary to see another person's perspective along with your own. Then you can encourage them to agree to a solution that is beneficial to you.

In the 1950s two scientists, James Olds and Peter Milner, were experimenting with rats to see whether they could teach them to do simple tasks like pressing a lever. They discovered that if you put an implant into a certain part of a rat's brain – the hypothalamus – and then apply a weak electric current, the rats actually like it, or rather they love it; if they were provided with a lever by which they could control the electric current themselves, the rats would press the lever about 2,000 times an hour, for hours at a time. If the rats were given the choice between feeding or pressing the lever, they would starve themselves rather than miss out on the current. This only worked if the current was applied to the part of the brain associated with dopamine release; if the rats were given a drug which blocked the action of dopamine they lost all interest in the lever. The reward system in a rat's brain had been discovered. It is assumed that humans have a similar system. Could brain stimulation explain the joyful feelings that the Russian novelist Dostoevsky said he felt preceding his attacks of epilepsy? He described it as 'a feeling of happiness such as it is quite impossible to imagine in a normal state and which other people have no idea of'.

In a more unpleasant experiment an electrified grid was used which gave a shock so painful to a rat's feet that if food were placed on the other side of the grid a rat would die of starvation rather than go through the agony of crossing. However, the rats would cross the grid to get to the dopamine pedal. It seems they were prepared to do anything to get that rush.

Drugs such as cocaine, nicotine, cannabis and amphetamine all raise the concentrations of dopamine in the brain, either by increasing the amount released or by blocking the mechanism that reabsorbs the dopamine, limiting its effects. Either way the person ends up with more dopamine. It's the same process as the one that happens when the rats stimulate their hypothalamus. The more addictive drugs – cocaine and heroin – release more dopamine.

It has now been discovered that even the anticipation of taking a drug can release dopamine. The problem is that in addiction this anticipation is experienced as a craving, which, despite the release of dopamine, isn't pleasurable. The power of dopamine could explain why addicts will sometimes risk losing their jobs and homes, even their families, in order to get another fix. As the brain adjusts to the new dopamine levels, the drug is needed purely to continue functioning. This is known as the dopamine theory of addiction, but what it can't explain is why so few people become addicted. Only 10% of people who use cocaine become hooked. Likewise only 10% of the American soldiers who regularly took heroin in Vietnam took it again once they were back home. This implies that a person's situation plays a big part in addiction.

A Canadian psychologist Bruce Alexander has lent weight to this idea with his creation of a luxury rat park. He trains rats to become addicted to a dopamine pedal, but then instead of placing the pedal in a bare cage, he installed it in the luxury rat park where there are jogger wheels, plants, warm nests, nice food, plenty of space to run around and even mountain scenery and streams painted on the walls. Once rats were in here and were occupied and presumably more contented they barely touched the dopamine pedal, in contrast to the caged rats so desperate to get to the pedal that they would starve themselves in order to reach it. This could provide part of the explanation for the fact that so many people who try drugs don't get hooked. Provided people have alternative stimulation in life they might not become addicted.

Another possibility is that some individuals have fewer dopamine receptors than others. With fewer receptors you need to take more drugs to derive the same pleasure as someone else. The level of addiction in rats has been successfully altered through manipulation of the number of dopamine receptors.

More radically, there is a suggestion that addicts are born with a gene which stops them from experiencing joy in the same way as other people. The idea is that some people are born with a difference in one particular form of the dopamine receptor gene. The gene controls the way that dopamine is released into the brain and comes in two forms, a common type and a rarer version known as A1. A team from UCLA in the United States led by Ernie Noble found that the same gene seems to be implicated in addiction to cocaine, heroin and nicotine, and even eating disorders. He believes that those with the A1 form of this gene don't feel the same joy from something like sitting on a beach watching a sunset or seeing a live concert, however brilliant. By taking drugs or drinking alcohol they can bring their dopamine up to the same levels as everybody else. This new pleasure leads them to become addicted to the experience. Not all addicts have the gene, but those who do find it harder to overcome their addiction. Even with smoking, people who repeatedly attempted to quit without success were found to be more likely to have the A1 form of the gene. With a team in Australia Noble tested his theory during the treatment of alcoholics. Patients were either given a placebo or a drug called bromocryptine which activates your dopamine receptors. They found that with the people possessing the A1 form of the gene the bromocryptine reduced their cravings almost completely. Moreover, they felt less anxious and remained in the treatment programme for longer than those who had the more common form of the gene. Once the bromocryptine had rectified their dopamine levels, the desire for alcohol faded. This gene won't explain all cases of addiction; Noble believes that about half of drug addicts might have it, suggesting that they might benefit from a pharmacological approach to treatment, while others might find counselling more effective. He hopes that in the future treatments for addiction could be targeted according to a person's genes. According to his theory as many as 30% of the population might have the gene, preventing them from experiencing joy. This, he thinks, might lead some people to seek risky pursuits, in the hope of feeling something.

If this is an accurate assessment then it is cause for some concern, but not everyone agrees with Noble's theory. Some attempts to replicate his results have ended in failure. It might be an oversimplification to suggest that just one gene is involved; there could be lots of different genes contributing – either in relation to a specific drug or general addictive behaviour.

The curious thing about joy is that if you ask people whether they would like to have the human equivalent of the rats' joy lever, as the Harvard philosopher Robert Nozick did in a thought experiment in 1989, most people say no. He asked them to imagine a machine which can produce whatever set of feelings you desire. You can experience success, pleasure or friendship in any combination and for as long or as short a time as you want to. The only catch is that although you would remain healthy you would have to spend the rest of your life attached to the machine. Despite the guarantee of feeling good for the rest of their lives, coupled with variety to dispense with boredom, people still say no, believing that however the machine made them feel, they wouldn't be truly happy.

the joy of exercise

When James was sixteen years old he took part in a cross-country championship. Near the end of the race he was in fourth place, but knew that he was too tired to make the pace needed for him to win. Then something strange happened. 'I was feeling really exhausted, but in a space of two minutes I moved into this rhythm and this beat where I felt absolutely invincible. I ran forward, took over first position and sprinted to the finish. I remember at the time I couldn't talk to anybody; I just had to be on my own. There was just this sense of something in my body rewarding me – almost a chemical-like feeling of power.'

James had experienced something which many athletes find elusive – an exercise high. Some people exercise vigorously for years without achieving even one, but when it happens the theory is that beta-endorphins are key. Endorphin is short for endogenous morphine, i.e. morphine made by your own body. They are the body's natural tranquillisers, which we release when we are in pain. Like morphine, they can also cause pleasure; the pain remains, but you don't care anymore.

Is it possible that these endorphins could actually bring feelings of joy? There's a physiological mechanism called the blood–brain barrier that protects the central nervous system by preventing most substances from crossing from the bloodstream into the brain tissues. The problem is that it's not known whether the endorphins released into the bloodstream during exercise can actually break through the blood–brain barrier in order to have their effect on the brain. Exercise highs are hard to study systematically because it's not an effect that can be easily quantified – who is to say what constitutes a high and what doesn't?

It does seem, however, that exercise can affect the brain, even if the results are somewhat milder than a full-blown high and this might enlighten our understanding of the way both our minds and the rest of our bodies influence our emotions. Two or three short sessions of exercise a week can make people feel demonstrably happier, particularly if they are depressed initially. Some GPs already prescribe exercise for patients with mild depression with some success. There have even been studies showing that exercise can be as effective as anti-depressant medication. As well as the suggestion that this is due to the release of beta-endorphins there is also a theory that, like recreational drugs, exercise might cause the release of dopamine, hence we feel good.

Although research into the effect of exercise on mood might sound straightforward it's not without its problems. For example, in some studies comparing one group who are prescribed exercise with a group not prescribed exercise, the people taking part were allowed to choose which group to be in. This inevitably introduces a bias – the people choosing exercise are likely to expect to see more positive results than those who avoid exercise. Studies have tended to allow a choice because it would be hard to persuade someone who had always detested exercise to take part in a study which might compel them to join a class. But if exercise could be shown to improve the mood of these people then it would be a powerful tool indeed. Yet another problem for the designers of these comparison studies is that factors other than the exercise itself might affect the group taking the classes. Participants might begin to feel better simply because they enjoy belonging to the group which goes to the special class. The fact that they are trying a new method of alleviating their depression might make them feel more hopeful and even the instructor's expectations that they feel better after a session could make a difference. Any of these could have a slight improvement on a person's mood before a single step of exercise has been done.

Despite the methodological difficulties there is some good evidence that exercise can make you feel good. The world's first Professor of Psychopharmacology, Hannah Steinberg, has suggested that the effect of exercise could be harnessed as a treatment for drug addiction. Just as addicts use methadone as a substitute for heroin, Steinberg believes the next step might be to use exercise as a substitute for methadone, helping the brain to provide its own alternative opiates. It's an intriguing idea, but she found it hard to persuade anyone to conduct trials.

A Danish study of just eight people addicted to drugs or alcohol found that exercise did seem to help them resist their addiction while they remained in hospital, but once they left they dropped the exercise programme and five out of the eight resumed their addiction. This study is so small of course that there could be other factors at play. The problem with using exercise in a deliberate attempt to induce feelings of euphoria is that even in experienced runners a high isn't achieved every time. There's also the question of whether the intensity of high could ever compare to that induced by drugs. Even if it worked on some occasions, however, it might be useful as one part of a treatment programme, with the added advantage that exercise is free and healthy.

However, it still isn't clear whether improvements in mood after exercise are caused by the release of endorphins. People feel better after exercise even if they have been given a drug which blocks the production of beta-endorphins altogether, suggesting either that other neurotransmitters might take the place of endorphins when they are blocked or that there are other factors at work. A person might feel better after an exercise class because they are pleased with themselves for having made the effort, or perhaps they enjoy seeing themselves become fitter, or simply feel satisfied at having mastered a new routine. These could all feed through to a person's self-esteem.

Finally there is the social interaction. If a person is at home feeling depressed, then visiting the gym with other people might provide a break from that isolation. If you use a drug to block the production of beta-endorphins, the improvement in mood after exercise still occurs, suggesting that endorphins might have nothing to do with the process.

In the Department of Kinesiology at University of Wisconsin-Madison there are treadmills, weight machines, exercise bikes, free weights and a swimming pool. While human guinea pigs pump iron every flutter of their heart can be measured. Outside there is even an arboretum where volunteers cut down the undergrowth while wearing transmitters which allow the scientists indoors to measure their heart rates. After they've finished the work, their levels of anxiety and depression are measured. With the help of this equipment and a team of volunteers, Professor William Morgan has developed a theory which might explain why exercise can make you feel happy – he calls it the distraction hypothesis. It is appealingly simple; exercise makes us feel good merely because it distracts us from the worries of the day. He found that although people do experience a decrease in anxiety as a result of vigorous physical activity, the effect soon wears off and within twenty-four hours you're back to where you started. Therefore regular exercise could be a way of topping up your joy and keeping a lid on anxiety. Too much exercise, however, can have the opposite effect. In experiments where swimmers progressively increased their training sessions from 3,000 metres a day to 12,000 metres a day, the athletes gradually became more depressed. Once they began to decrease the distance they swam, their mood slowly returned back to normal.

There is some more good news for people who prefer to sit still. Professor Morgan also found that when he asked people to sit in a quiet room in a comfortable old leather armchair or lazyboy, as he calls it, for the same length of time as an exercise session, they felt just as good afterwards.

A final theory of why exercise might make you feel good is the thermogenic hypothesis – the idea that you feel happy after exercise because your body temperature has increased and that it is this high temperature which is responsible for the release of beta-endorphins, which in turn make you feel good. Not surprisingly, research in this area began in steam baths and saunas in Scandinavia. In one study back in 1972 the volunteers who took part had a twenty-minute sauna followed by a ten-minute shower for which they were paid in cash and in beer. Before and after the sauna the mood of the volunteers was measured. Beforehand two of the twenty male volunteers had warned the experimenters that they didn't like saunas and indeed four people did find the sauna so stressful that they had to be let out early. However, overall people did feel less anxious after the sauna, but so did the control group who had undressed, sat waiting on a bench for twenty minutes and then had the ten-minute shower.

More recently William Morgan's team have tried to determine whether it's the exercise *per se* or the rise in temperature which makes people feel good after exercise. He found that if the whole body is heated mood can improve for up to twenty-four hours and in the bloodstream at least, levels of beta-endorphins rise. The team have come up with ingenious experiments in an attempt to establish the answer such as getting one group to exercise dressed in everyday fitness gear while another group exercised clad in warm clothes, a hood, gloves, a surgical mask and two blankets. Despite the rise in body temperature, not surprisingly the hot group ended up feeling more anxious than the others, rather than less. Another approach is to prevent the exercise from increasing the body's temperature by lying in a cold bath for half an hour before exercising. It was found that if a person's body had been cooled beforehand their temperature remained low and they felt no better after exercise, whereas the people whose bodies had not been cooled did feel happier after exercise, suggesting that it could be the rise in temperature, not the exercise itself, that makes people feel better.

Another way of testing this theory is to prevent body temperature from rising with exercise by doing that exercise under water, but when this was tried with scuba divers they still felt better afterwards, suggesting that it's not the warmth that's doing the trick. At the University of California Shawn Youngstedt and his team put volunteers on exercise bikes under water with their heads poking out above the surface. There was no improvement in mood after exercising, but the researchers do warn that this might have been due to the novelty of the task or anxiety caused by having their oesophageal temperature taken – a procedure involving the insertion of a tube up the nose and down into the throat. Other experiments have involved exercising while wearing a scarf filled with ice and even having a rectal temperature taken before, after and (somehow) during exercise.

In this study anxiety did decrease after the exercise session, but perhaps the participants were simply relieved that it was over. So it seems that exercise can make people feel good, but we're still a long way from knowing exactly why.

smiling

'Cheer up! It might never happen,' builders like to shout as I walk along the street perfectly happily, not feeling sad, but daydreaming. Irritating though these remarks are, the builders might in fact be on to something. While it makes intuitive sense that exercise might make you feel good, the next topic is more primitive and rather more surprising. Remarkable research has found that we can influence our brain chemistry through something far less taxing than exercise – smiling.

The first person to make a serious study of the smile was the French neurologist Duchenne de Boulogne who determined which muscles were needed to make each facial expression by stimulating each muscle with an electric current. In 1862 he published a book chronicling his findings which included photographs of a man clearly in a state of terror while metal implements were held against his skin. Fortunately for the man, the expression was created purely using the electric current and he could feel no pain because he had a condition which rendered his face numb. Using this method Duchenne studied in detail the muscles responsible for different facial expressions, including smiling. He soon identified the difference between a genuine and a fake smile; the key is a muscle above the eye where crow's feet form, called the *orbicularis oculi*. Anyone can fake a smile by raising the corners of their mouth and baring their teeth, but if the muscle above the eye remains motionless the smile looks wooden.

Today the world authority on facial expression is Paul Ekman, who uses his expertise to, amongst other things, train police to spot when people are lying. He's found that a smile is hard to fake because most people can only voluntarily contract one part of the *orbicularis oculi* – the inner part which tightens the eyelids. Only 10% of people can decide to contract the outer part of the muscle which pulls the eyebrow down at the same time as raising the cheek and pulling up the skin below the eye. Therefore the way to judge whether a smile is genuine is to see whether the cheeks move higher and the eyebrows tip down slightly. You can experiment by standing in front of a mirror faking smiles. Hopefully you will feel so daft that eventually you smile genuinely and then you can see the difference. Ekman's other clues to spotting a false smile are that it can be asymmetrical, with a timing that isn't quite right – either just too early or slightly too late. If the person faking the smile is right-handed, the left-hand side of their mouth will tend to move up more.

The authenticity of a smile can reveal the unexpected. In one study women's smiles were analysed from a college yearbook. Thirty years on the women with the genuinely happy smiles using the muscle around the eye, were more likely to be married and happy. The researchers did rate each woman's looks in case the smiley women were simply the prettiest and perhaps therefore the most likely to have found partners, but this wasn't the case.

It's not only adults who can fake a smile. Paul Ekman discovered that if a stranger approaches a ten-month-old baby, the baby might well smile, but that smile won't involve the crucial eye muscle, but if their mother approaches them, it does. Although the baby isn't deliberately faking a smile, at this young age it can already smile politely.

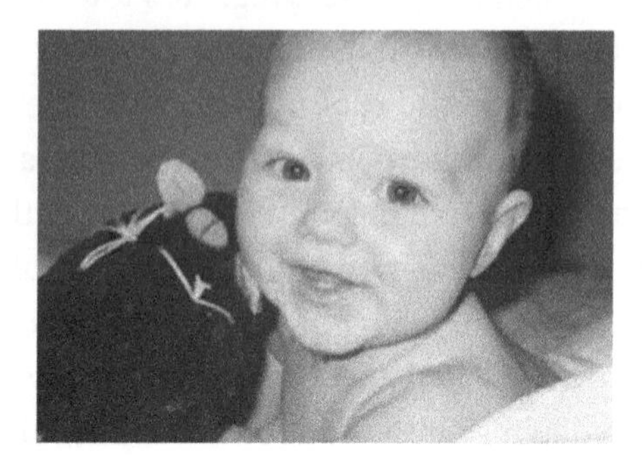

From the age of four or five weeks babies smile at any human face which nods about two feet away from their face. They seem to want to smile, to engage in communication. This helps the baby and parents to develop a strong bond and is one of the first rewards that exhausted new parents receive for all their hard work. Smiles in very young babies used to be dismissed as wind, but using a new scanning technique developed in London, babies have been spotted smiling in the womb. We can't prove of course that the baby smiling in the womb is expressing joy but perhaps conditions in the womb vary enough for the baby to feel more comfortable on some days than on others. This has blown a hole in the myth that babies only smile through imitation, although it is true that the more you smile at babies, the more they smile back, just as they will copy other expressions like sticking out their tongues. However, they are born with the ability to smile. In the past it was assumed that babies could experience very few feelings – not even pain – hence the absence of anaesthetics for young babies at one time. Now it is accepted that even a very small baby might be feeling happy.

Joanna Hawthorne, a research psychologist at Cambridge University, works with new parents, encouraging them to judge their babies' states so that they can choose the right moment for interaction – when the baby is alert but neither too hungry nor too full. It may only last a few seconds but this is the time when an adult and a small baby can take turns in smiling. If a baby smiles at us, we take it as a sign that they like us and we smile and behave warmly back and so the cycle continues.

Smiling also appears to play a crucial role in social interaction between adults. There is a rare condition called Moebius syndrome, where a person's face becomes paralysed, leaving them unable to smile. One consequence is that they can often find it hard to make or keep friends. This suggests that there is an important social element to smiling, as does the fact that however happy people are with their own company, they smile far more when others are present. In an experiment conducted in a bowling alley it was found that after achieving a strike, people beamed more when they turned to face their friends than at that most satisfying of moments when they watched all the triangle of skittles collapse in a heap.

More recently some Spanish researchers took advantage

of the location of the 1992 Olympics to watch twenty-two gold medal winners very carefully, including Sally Gunnell. They observed them while waiting to mount the podium, standing on the podium and turning towards the flagpole while their national anthem was played. They found that people smiled more during the second stage, despite presumably feeling happy that they'd won in the other two stages as well. This the researchers took as a demonstration of the fact that people are more likely to smile when they're in a social situation rather than when they're just feeling happy. However, it should be remembered that there are social rules prescribing when it is and isn't acceptable to smile; one moment when Olympic winners are expected to look sombre is during their national anthem.

The mysterious relationship between smiles and joy has been investigated by Paul Ekman in an extraordinary experiment. Without actually telling people to smile, he gave people precise instructions about which muscles to move, and despite being unaware that they were smiling, the physical process of moving those specific muscles into a smile made people feel happier. In a variation on this experiment the psychologist Fritz Strack gave people a pen to hold in their mouth. The instructed action of holding the pen between the teeth without touching the lips mimicked the movement of a smile, but once again the subjects of the experiment were unaware of the expression they were making.

They were given cartoons to watch and rated them as funnier when the pen was in this position than when they were told to suck the pen with their lips closed around it.

The idea is that the facial muscles are so sensitive that they can feed back their position to the brain and somehow lift the smiler's mood. This is known as the facial feedback hypothesis. It seems that what we previously thought of as the result of an emotion (a smile) can also be the cause. Even posture can change your mood; sitting upright makes you feel happier than when you're slumped. It isn't clear exactly how these processes might work and, indeed, the change in mood usually doesn't last for long. Is it the case that you become aware that you are smiling which makes you feel good or is it purely physiological?

The nineteenth-century American philosopher and psychologist William James believed that emotions are caused by our awareness of physical sensations and claimed to have used his own theory to work his way out of depression; the more he smiled, the better he felt. Perhaps it is the case that smiling bravely through your tears can make you feel better, in the same way that looking up at the sky if you feel miserable improves your mood slightly, because it's a movement you would usually make when you are happy. An extension of this theory might explain why we tend to like smiley people. The more they smile, the more we smile back, which could, in turn, make us feel happier.

an experiment

Try this. Stand up and then move your body as though you are laughing uncontrollably. Bend forward, clutching your sides, with your shoulders shaking. You could even introduce the occasional knee-slap. Now try to remember exactly what this feels like physically within your body. Then stand up straight again and imagine that you're doubled up with laughter, but this time don't move, just think of those movements. Don't think of a funny occasion or a joke. Simply imagine that your body is experiencing uncontrollable laughter and then see how you feel.

This exercise isn't easy, but Nakia Gordon from Michigan State University succeeded in training people to do just this. After three one-hour training sessions, along with plenty of practice at home, most of the volunteers in her studies were able to lie absolutely still in a brain scanner while imagining either laughing, crying or walking. Each movement resulted in a different pattern of brain activation. This experiment shows that emotions can be linked to bodily movements, something William James had thought so long ago. Moreover the people taking part felt happier after imagining laughter, and sadder after imagining crying, even though they were instructed only to imagine the movements, not the feeling of those emotions. This opens up the possibility that it's not just that smiling can make you feel better, just thinking about smiling or laughing might do the same trick.

the laughing umbrella

On the pavement beside a main road in south London two pairs of legs are visible emerging from a huge yellow tent which seems to hover above the people's heads and only comes down as far as their knees. The whole tent is shaking slightly and from inside there's the sound of a person laugh-

ing uncontrollably. A few people stop and stare, wondering just who these two people are and what could be going on. Then the tent is lifted off and collapsed like an umbrella. Red-faced with laughter the man thanks the woman with blonde, spiky hair for cheering up his day and walks off up the road, still chuckling quietly to himself. Then the woman approaches another passer-by, introduces herself as the artist Nicola Green, and invites her to come under her umbrella. She is collecting laughter and has found that her customised umbrella provides people with the privacy to laugh while she stands with them making a recording. As a portrait artist she had explored unusual ways of capturing the 'essence of a person' by painting a family portrait as the backs of people's heads or a line of feet. Then she noticed how much a person's laugh reveals and decided that this was the way to access their joy.

Laughing on cue isn't easy, as I discovered when I stood in her living room under her laughing umbrella. If you're passing by in the street then the ludicrous nature of the situation might make it easy, but since I'd arranged to meet Nicola and knew exactly what she was doing, when it actually came to it, I found it surprisingly difficult to laugh. Luckily she has ways of helping reluctant laughers. By encouraging you to fake a ridiculous cackle, she makes you laugh genuinely. She even went on the tube in London, where laughing between strangers simply doesn't happen, chose a carriage and then shouted out, 'I want to record your laughter. When I blow the whistle I want you all to laugh.' Twitching glances crossed the carriage between passengers. Should they do it or not? Could you risk being the only person to laugh? Would it be better to ignore her and hope she moves on? In that split second people make a decision about whether to laugh. She found that either a whole carriage would laugh or nobody would. At Victoria station she somehow persuaded the platform attendants to make a tannoy announcement asking everyone to laugh. Eventually she made her laughter collection into a record, encompassing everyone from her dentist to a vicar in Covent Garden.

It struck me that people often thanked her afterwards for the fun they had had. They clearly felt good, suggesting laughter might provide some sort of physical release. Usually laughing does make people feel better, hence the rows of people happy to sit in laughter clinics, chuckling contentedly. Occasionally, however, laughter does not denote happiness.

the laughter epidemic

In 1962 a very strange thing happened in the village of Kashasha near Lake Victoria in what is now Tanzania. At a small Catholic girls' boarding school there were 159 pupils. On 30th January three girls began laughing for hours at a time, then crying and then laughing again. Two weeks later they were still laughing. By the middle of March so many pupils were laughing that the school was temporarily closed down. No one knew the reason for the laughter or the tears, but it seemed to be contagious. After the school was closed the girls went to other schools and the epidemic seemed to spread. Soon boys were laughing too. In the village of

Nshamba the hysteria spread to more than 200 inhabitants and even to young adults. Soon two boys' schools had to be closed down for the same reason. Blood tests were taken to test for food poisoning, but everything appeared normal. Six months later people were still giggling – it was a laughter epidemic. The outbreak is often described in works on laughter, usually as a joyful occasion. But it seems it was no laughing matter. Christian Hempelmann from Purdue University in the United States believes that in fact the girls may have been suffering from a mass psychogenic illness of which laughter was one of the symptoms. He believes the condition was caused by the transition occurring in the country at the time. Independence had been achieved in December 1961 and schools were growing with the aim of providing free education for as many children as possible. Hempelmann believes the hysteria might have been a response to the stress of suddenly going to Western-style schools; far from enjoying themselves the pupils were suffering.

Although the researchers who first published details of the case in *The Central African Journal of Medicine* in 1963 themselves suggested that mass hysteria could be the explanation, it's intriguing that a humorous explanation has been sought ever since, with writers wondering what the original joke could have been, a joke so good that it apparently made people laugh for months. We seem to have a desire to make sense of our emotions, particularly when they are as intense as uncontrollable laughter. The need to look for explanations was illustrated by the case of a sixteen-year-old girl who was examined in 1998 as part of an investigation into her epileptic seizures. When the doctors applied an electric current to a certain part of her brain she burst out laughing, even when distracted by other activities. It was intriguing that each time she looked for a justification for her laughter, saying how funny the doctors were or that the picture of the horse they had shown her was hilarious. Although joy can be free-floating, when we observe that joy in ourselves we are inclined to look for meaning. Instead of searching for an explanation for individual instances of happiness, perhaps the crucial question is why we have evolved to experience this feeling at all.

the purpose of joy

Everything had been planned for the Saturday night. We were staying with friends in Scotland, the wine was chilling and the lamb was roasting. Everything was in place for us to have an enjoyable evening. Then the host's mobile phone rang yet again. He was a vet on-call and throughout the weekend we'd become accustomed to his answers which usually seemed to be, 'It sounds as though your dog has a cough. I don't think an emergency appointment will really be necessary.' We assumed this call would be more of the same, but it wasn't. He rushed out to the surgery at the end of the road, with his wife as assistant. Our main job was to keep an eye on the roast potatoes or so we thought until the phone rang half an hour later. They needed more pairs of hands and we would be able to get there faster than the veterinary nurse. We let ourselves into the surgery, found the operating theatre in the basement and there on the table was a tiny Chihuahua with a transparent tube in her mouth and her insides spilling out of a red slit across her stomach. She was having a Caesarean section. The vet slid his hand inside and swiftly pulled out something that looked like a dead, shaved mouse. As he shook it hard, its head flopped backwards and forwards. To us, with no knowledge of veterinary medicine, it looked as though this might break its neck. Instead it came to life, still floppy with eyes still tightly shut, but coughing softly. Meanwhile we followed his instructions to create a miniature hot water bottle by filling a plastic glove with warm water and took it in turns to rub the tiny creature's chest continuously to keep it breathing. It was hard work and to us the puppy soon looked dead again, but eventually he could breathe alone. We were ecstatic. By now the mother was coming round. 'Let's show her the puppy. Can we put it beside her?' we asked. 'You can if you want to, but she won't care. She's in shock.' She woke up, shivering and horrified. She wasn't quite as thrilled as we were. Unlike the proposed entertainment of a delicious meal and nice wine with our friends, this joy was totally unplanned, but all the better for it.

The fact that unexpected variety can bring such pleasure could indicate one of the purposes of joy – to motivate us to experiment with different activities or places. When we feel scared we are forced to narrow our focus in order to concentrate on the danger, but the opposite happens when we feel happy; our perspective broadens, allowing us to make new discoveries, a skill which may not be essential today, but which would have been useful in hunter-gatherer communities. Experiences of joy also strengthen ties with other people, ties which could be important for survival.

Dylan Evans, an expert in artificial intelligence from the University of the West of England, believes that joy could have a further evolutionary purpose; joy advertises our mental and physical fitness. If you are able to pursue joy openly then you must have met your basic needs such as food and shelter, which makes you a more attractive mate.

Some researchers believe that we have a basic brain system for joy, which explains our tendency to take any opportunity to be playful. If you watch people working in an office, provided they have the time and the autonomy, they are ready to take any opportunity to feel joy; they are primed to have fun. I remember a colleague once admitting he could do a one-handed handspring. Then someone else said they could do that dance move from *Singing in the Rain* where the dancer steps up onto a chair back and lets it tip over backwards until it brings them to the floor. Inevitably the rest of us insisted on proof and we soon discovered that office ceiling fans make handsprings tricky and why actors in musicals avoid using typists' chairs on castors. When I was a child I remember being very shocked when my father broke his finger in a desk-jumping competition in his office – shocked not that he'd injured himself, but that adults would play in that way.

The first person to win the $100,000 Templeton Positive Psychology prize was Barbara Fredrickson from the University of Michigan. After studying positive emotions for more than a dozen years, she believes that they have a specific role – to ameliorate the effects of negative emotions on the body. To demonstrate this theory Professor Fredrickson first induced anxiety in a group of volunteers by telling them that in one minute's time they would be required to give a speech which was to be filmed for evaluation by the rest of the group. The nerves soon began to show in their raised blood pressure and heart rates. Then they were given a film to watch which was either funny, happy, sad or neutral. It was found that the people who watched either the funny or happy film recovered their normal heart rates and blood pressure faster than the others, indicating that positive emotions can help to undo the cardiovascular effects of negative emotions. This might explain why certain stressful professions such as medicine become known for the dark humour that often accompanies the work. Medical students working in casualty for the first time are far more likely to recount tales of patients who claim to have fallen onto a peeled carrot while gardening than to concentrate on grisly stories about the appalling injuries they have seen. The deliberate creation of joy through humour works as an effective coping mechanism.

The thought of future joy also has the benefit of encouraging us to plan ahead. One of the happiest moments in my life was the day my best friend Jo passed her driving test. We were seventeen and lived about twenty miles apart, but we knew that her ability to drive would lead to multiple opportunities for potential joy. No longer would we have to rely on our parents or older boys for lifts. To celebrate we set off in her rusty, royal blue Chevette called Cyril, which although old and slow had that essential element – a tape player. With the windows down we sang along to U2 as loudly as we could while driving around the county paying people surprise visits. Despite the fact that we were celebrating a joy and freedom that was yet to come, we felt ecstatic. This ability to imagine happiness in the future allows us to make plans which we hope will end in joy, even if it is necessary to forfeit some fun today in order to achieve them.

There is a children's story in which the main character is given a choice of gifts – she can have a delicious chocolate fudge cake or the recipe for the cake. The moral of the story is that the wise decision is to go for the recipe because then you can make endless chocolate cakes in the future – in other words, delayed gratification. However, we do soon learn that by forgoing a little joy now – by getting out of a warm bed to go to work, for example – we can experience even more joy later – when we have earned enough to go on holiday.

There is an intriguing theory developed by Michael Apter from Georgetown University in the United States. He proposes that we alternate between two states – the 'telic' where we make the effort to pursue serious goals and the 'paratelic' where we seek out experiences for their own sake, often in a playful way, regardless of the future. His 'reversal' theory suggests that we all switch back and forth between these states, depending on where we are and how we feel. So perhaps the answer to the children's quandary about the chocolate cake is that your choice of the cake or the recipe depends on whether you are in a telic or paratelic state. Presumably when my father broke his finger he was in the latter.

the route to a happier life?

While research on positive emotions has increased in recent years and the subject has begun to be taken more seriously, the question remains of whether research on happiness can in reality help us to experience this emotion more often. It's not hard to improve mood in a laboratory. Show people a film of penguins waddling, playing and sliding on the ice and soon everyone feels a bit better. Lying in a flotation tank, singing out loud or listening to stirring music can all lift mood temporarily, but the late social psychologist Michael Argyle wanted to know what gives people lasting happiness. After years spent researching the subject he concluded that the answers lay in attending church, joining sports clubs and watching soap operas. His analyses showed that on average the people who did these activities were the happiest. All three pastimes share both a sense of belonging and of social occasion; even identifying with the characters in a soap opera can induce feelings of belonging in a way that most other TV programmes can't. In fact, Argyle found that people who watched a lot of television were more unhappy than average. This could be a reflection of the situation which led them to have so much time to spare at home; maybe they were more likely to be unemployed, isolated or unable to afford to go out. Although the people attending church or sports clubs were the happiest, they had of course chosen to join these groups. The research doesn't demonstrate that if everyone joined they would be happier. If you like neither sport nor religion it could make you more unhappy. When I met Michael Argyle a couple of years before he died his tip for a happier life was simple – to go for brisk ten-minute walk twice a day. People reported feeling better for two hours afterwards, so potentially two walks a day could result in four hours of uplifted mood.

The simple idea of planning activities with the aim of improving mood has been used therapeutically. Unhappy people were asked to keep daily records of both their activities and their moods. Then the lists were analysed to find out what made them happiest and these activities were then incorporated more frequently into their routine. It transpired that people weren't in fact doing their favourite activities very often, but when they did they felt happier.

However, unfortunately for people who are unhappy, Michael Argyle found that people were relatively stable in their degree of happiness, regardless of what happened to them. A happy person suffering a misfortune might feel temporarily less happy than they were before, but that could still be happier than many people feel despite their lives apparently having gone well. Finding it easy to get on with people and taking exercise were also strong predictors of people's happiness levels. As for money, becoming richer only makes people happier if the amounts are substantially higher than their expectations of earnings. With all these studies, however, there is the issue of exactly what people mean by happiness when they are filling out a questionnaire. Are they talking about moments of joy or an overall state of happiness? Why is it, for example, that five times as many Norwegians claim to be happy than Italians? Can the differences really be that extreme or are people interpreting the questions differently? Some researchers try looking at the reverse. Rather than measuring how happy people claim to be, they look to see which countries have the fewest suicides, but of course a low suicide rate only tells us that few people are suicidal, not that everybody else is happy.

Joy is perhaps a rather unfair emotion because it is biased towards those who already feel happy. Happy people are better at coping with distressing events; they are more attractive to others; and they find it easy to conjure up happy memories. Moreover, because they feel happy they smile more, sending them on an upward spiral of happiness which, through facial feedback, might reinforce their good mood every time they smile. Meanwhile those who could benefit the most from that extra fillip granted by smiling, aren't smiling because they feel sad, left on a downward spiral with unhappy memories foremost in their minds. From this perspective it's hard to see how sadness could possibly be functional, but as we shall see in the next chapter even sadness has its place.

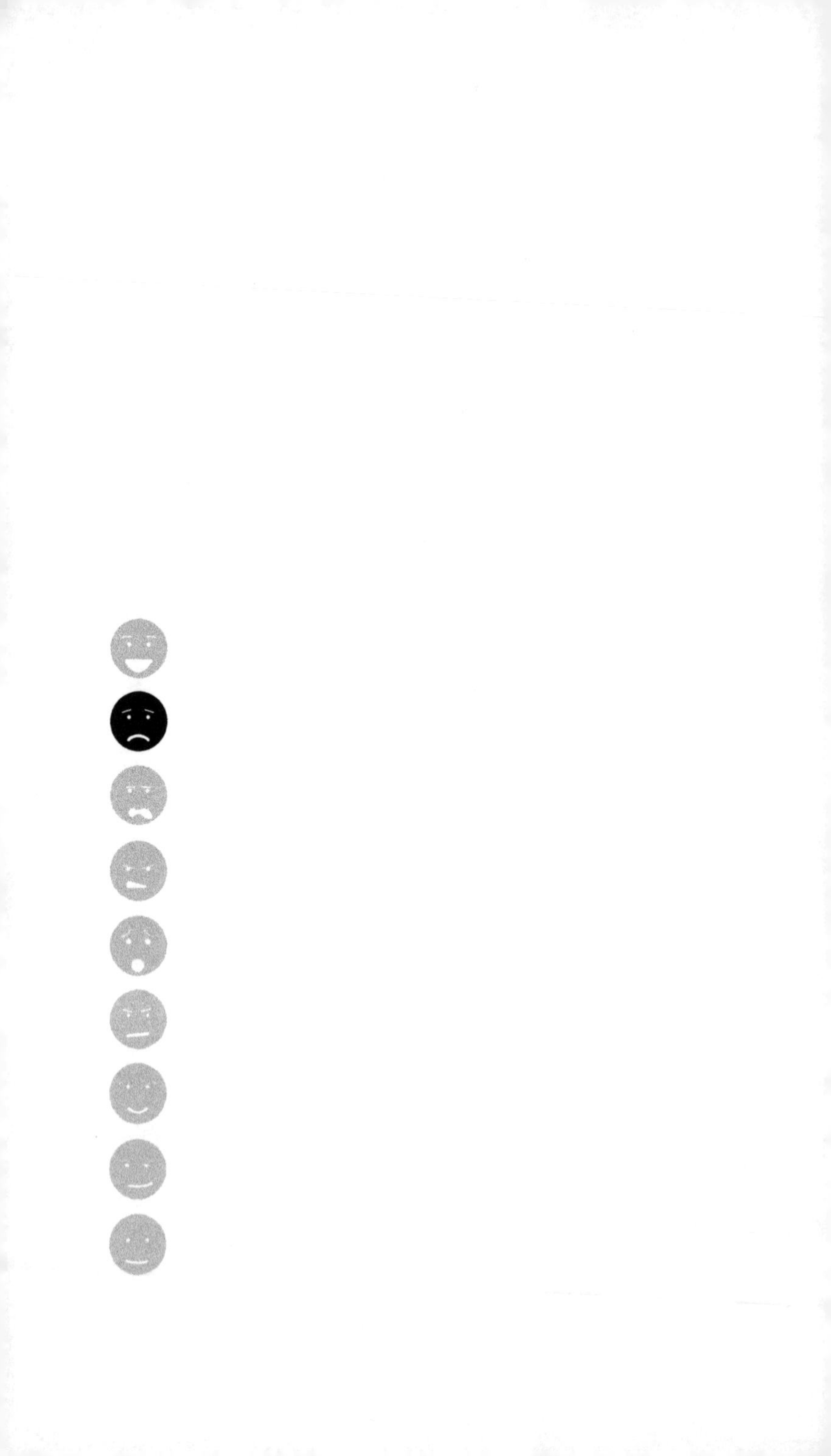

two

sadness

Airports are not supposed to be sad places. People are off on 'the holiday of a lifetime', 'a dream honeymoon' or a gap year 'finding themselves'. Wherever I'm heading I find myself looking up at the departure screens, imagining which flight I'd board if I could choose: will it be Kuala Lumpur or Moscow, Lima or Nairobi? But this is just my experience. Airports aren't always happy places.

A woman walks slowly towards the automatic glass doors carrying a soft black bag on her shoulder and a linen jacket wedged in the crook of her elbow. She looks back, slightly embarrassed, aware that others are watching. Lifting her arm, she waves hesitantly at her family who stand with their mouths turned down. Their eyes are already wet. She shows her boarding pass to the man at the desk and walks through the doors ready to snake her way to the x-ray machines. Her family wait for the final glimpse, then put their arms around each other in comfort. For a while they stand still. How long should they wait? They turn and head slowly for the car park, not noticing a young man watching grimly as his pretty girlfriend passes through the same doorway. Will she still be his after a year's backpacking? A party of ten or twelve are seeing off an elderly Indian woman carrying two vast, checked shopping bags. The toddler is busy running straight into the stretchy barrier and bouncing back off it, but everyone else in the group is crying. Will this be the last time they see her? The stories of all these visitors to the departure lounge are very different, but they're all sharing the same emotion – sadness.

When someone else read this description they said that imagining those people made them feel sad, but in fact I'd made these stories up. Nevertheless my friend was experiencing that strangely enjoyable sadness you get if you watch a sentimental film. The next time I was at an airport, instead of pondering where to go, I looked around for the sadness and there was plenty. This time it was genuine.

Real sadness, imagined sadness and a slightly luxurious, voyeuristic sadness – this is an emotion which can work at different levels. A little sadness can be quite pleasurable. If you look back at happy times, you might feel sad that a part of life is over, never to be retrieved, but at the same time it's a warm, nostalgic sadness, free from despair. Melancholy music can have the same effect and we choose to listen to it specifically because it prompts an indulgently enjoyable sadness. In fact over the centuries sadness hasn't always been seen as a negative emotion. In her work on diaries from seventeenth-century England the researcher Carol Barr-Zisowitz found pride in the feeling of sadness. It was even considered to be the opposite of sinfulness; if you were sad you were seen as patient and wise, despite your difficulties. She also notes that in some societies sadness can have the same effect today. In Iran and Sri Lanka for example a degree of melancholy is taken as an indication of a person's depth.

Even in cultures where sadness is on the whole perceived to be negative, the absence of this feeling can be seen as problematic. A life spent feeling ceaselessly happy due to the drug soma was part of Aldous Huxley's nightmare vision of the future in *Brave New World*. You could argue that if you never felt miserable then neither could you ever feel true happiness. However, we wouldn't consider a life without illness to be a problem, despite a lack of contrast with good health. In fact when you feel ill the idea of feeling well again seems blissful, but after only a day or two of good health it's easy to forget how good you feel. Just as it's hard to appreciate every moment of well-being, it is hard to relish the absence of sadness.

In this chapter I'll be exploring why a sad face expresses so much and what happens in the brain when we're feeling sad, as well as the chemical secrets held within that strange symptom of sadness, crying. First it is necessary to understand the purpose of sadness and its close but more complex relation, depression.

the purpose of sadness

'You just can't imagine ever feeling happy again. You feel so helpless. There's just a big void. You can't even remember what it's like to feel happy. At first you don't know what's going on. You don't know why you're crying all the time. You go onto a different level. You can laugh with people, but it's a superficial laughter that starts in your head and only goes down to your neck.'

When I knew Chloe at school she had always been happy, pretty and popular, but when she was in her early thirties the end of a relationship was followed by a particularly stressful period at work. Suddenly she found herself bursting into tears without warning. 'When it first happens it hits really hard. You forget about food. It's too much effort. I remember being upset because my room was untidy but I just didn't have the energy to tidy it. I phoned my mum and she said she'd come and help me. My bedroom was right next to the front door, but just getting up to let her in made me feel so tired that I had to go straight back to bed. It was very different from just feeling sad, but I think only people who've been through it can understand that. Other people think you could do something constructive, if only you tried, but it's impossible. You feel such a failure and so guilty for not being able to do anything. You just feel useless and you take everything personally. You feel a complete burden, but you can't even explain it to people properly.'

When depression can become this painful, it raises the question of why sadness evolved at all. Although we vary in our tendency to feel sad, we all feel it, which suggests that it is either an accidental by-product of evolution or that it serves a distinct purpose. It does have to be remembered, however, that evolution takes no account of our pain; natural selection concerns the provision of a life where we can survive long enough to reproduce successfully. The purpose is not to furnish us with a life that's happy nor even healthy. Despite this, it is hard to see where sadness fits in. It paralyses people, preventing them from succeeding at work or finding partners, let alone reproducing. As we saw in the chapter on joy, a person who feels happy continues to pursue those activities which bring them joy, whether that entails remaining in a relationship or working hard at a job they like. Happiness or even the possibility of future happiness spurs us on. Sadness does the opposite. It can slow a person down to the extent where they stop working; they cease seeking other people's company; or they even stay in bed. People with clinical depression struggle to find the will to start anything new at what appears to be the exact time that extra energy is needed to make major life-changes. The cognitive scientist, Keith Oatley, believes that we experience strong emotions when we face a crossing point; their purpose is to act as a bridge to the next step in life. Whether or not we decide to change direction is irrelevant. Sadness concentrates the attention, forcing us to stop and take stock. The problem with this theory is that although people do slow down and isolate themselves when they're feeling sad, they don't necessarily spend a lot of time deep in self-examination. In one study people were asked how they tended to behave if they were feeling sad. The most popular answers were listening to music or taking a nap. Scrutinising one's life did not come high on the list.

Taking the idea even further, the American psychiatrist, Randolph Nesse, argues that the function of sadness is to control our energy levels. The idea is that if our chances of success on a project are low, on some level we realise this and start to feel miserable. Then we lose energy and motivation and abandon the project, which saves us from wasting time on something fruitless. It's even been argued that the treatment of depression with drugs might artificially rid us of a useful emotion.

In some ways this theory does sound plausible, but extreme depression can lead to suicide, hardly an ideal way for evolution to continue the species. In the UK and Republic of Ireland somebody commits suicide every eighty-two minutes. Moreover, not everyone who becomes depressed is following a course which is doomed to end in failure. Sometimes it is the onset of the depression itself which ruins a person's chances of success. Maybe it's the case that a little sadness can be useful, but for a few people the system reacts too strongly.

A more extreme suggestion is proposed by Anthony Stevens and John Price in their book *Evolutionary Psychiatry*. This is the idea that in hunter-gatherer communities depression was a useful protective mechanism. A person who stopped contributing to the community could be cast out and at that time ostracisation could well result in death. Instead people who feared they were of no use to their community would become depressed. They would then be considered ill and cared for rather than thrown out. In many societies today this is not quite so successful because the community might not rally round, leaving a person isolated. The problem with this theory is that it fails to explain why a person's view of themselves becomes so negative in the first place. Many people become depressed despite knowing they are valued by those around them.

An alternative approach to the purpose of sadness is to view it as an effective signal to others that you are in need of help. However, once again there are contradictions. At the very time they might benefit from the care of others, unhappy people often turn inwards, eschewing company. Moreover there isn't a universal response to another person's sadness. There's no guarantee that telling someone that you feel sad will bring you help. It will depend on both the individual and the culture. Anthropologists have attempted to look at attitudes towards sadness around the world, but once the word 'sadness' is translated into another language it might not have precisely the same meaning. One clever study overcame the translation problem through the use of photographs of facial expressions. People were asked to label the expressions in their own language and these were then translated into English by a native speaker. Through this method it was found that there are five types of sadness in the Greek language and six in Japanese. Therefore elements of sadness might be universal, but it might not be a single, basic emotion in all cultures. In some societies sadness is viewed as an acceptable emotion only for women or children. This happens to an extent in Western societies, where unhappy women are more likely to experience symptoms of depression, while men are more likely to express their misery through violence. If the purpose of sadness is to attract help, then it's a system lacking in efficiency because help is not always forthcoming. Many would prefer to spend an evening with somebody cheerful than listen to the woes of another.

So, we cannot be certain of the evolutionary purpose of sadness. It could involve elements of all these theories – an emotion which slows us down, gives us time to reflect and if necessary change plans, whilst signalling to others that we need them and thus strengthening those bonds. Certainly our faces are remarkably good at conveying our feelings when it comes to sadness.

the sad face

If sadness is to serve any communicative purpose, it needs to be obvious. When Darwin spent time studying the way the face expresses misery he noted that in many different cultures pulling down the corners of the mouth indicated sadness, hence the phrase 'down in the mouth'. What intrigues me about the facial expression of sadness is that it's usually more fleeting than the experience of sadness. You might feel sad for days, but only look sad some of the time. Concealing your feelings is possible, if tiring. As I discussed in the last chapter, fake smiles can be detected due to the lack of involvement of the muscles around the eyes. Darwin describes sitting opposite a woman in a railway carriage who looked perfectly content, but for the fact that the corners of her mouth were turned down. This revealed her true feelings. Even a slight turning down of the corners of the mouth conveys sadness. I've suffered from this myself. When my mouth is closed and not moving, there's a slight tendency for the corners to turn down and people often ask me whether I'm miserable when in fact I'm just concentrating. A few months ago I was standing waiting to meet a friend in Chinatown in London, feeling perfectly contented, when a policeman approached me to ask whether I was all right. He and his colleagues had seen me on the closed circuit camera screens in their van and thought I looked so sad and anxious that they wondered whether I was being followed. The same sort of thing used to happen to my grandfather, whose mouth was also wont to turn down.

In Tierra del Fuego local people tried to explain to Darwin that the captain of the boat was feeling sad by pulling down their cheeks with both hands to make the face as long as possible. Darwin went to great lengths to explain another feature of the expression of sadness which he called 'obliquity of the eyebrows'. The ends of the eyebrows nearest the nose are raised, causing the brow to furrow slightly. He calls this the grief muscle and notes that although it is contracted when we feel sad, only some people can move it voluntarily, an ability which seems to be inherited. The world-renowned expert on facial expressions, Paul Ekman, has studied identical twins who were raised apart and discovered that if one twin is able to flex this muscle, so can the other.

Along with general posture and tone of voice, facial expressions help us to spot emotions in others, but while some people are excellent at reading these signals and working out how another person is feeling, others won't even notice.

In fact the emotions we are best at observing in others are the same emotions we are experiencing ourselves. Therefore, although one might expect a depressed person to be too unhappy and concerned with their own misery to notice how someone else is feeling, in fact it's the reverse. Depressed people are the fastest to spot depression in others, while happy people are best at noticing the emotions of other happy people. It's almost as though the feelings we are currently experiencing ourselves are so strong in our consciousness that we find ourselves drawn to that feeling in others.

the sad mind and the sad body

Three years ago Julia wheeled her trolley past an empty customs desk. Her trip to Vietnam had been fantastic and as usual when she was tanned and rested, she felt great. Wouldn't it be nice if someone had come to meet her? To her surprise she saw that someone had. Her sister's boyfriend was leaning grimly on the barrier, but it wasn't nice at all. Why was he here? What had gone wrong? Had her mother's cancer come back? No, it wasn't that. He told her that everyone was safe, but that her parents' thatched cottage in Suffolk had burnt down. Julia was devastated. 'It was a 400-year-old cottage and my dad had lived there since he was ten. When I got back to my sister's house she showed me the front page of the local paper and there was a picture of my dad who's an artist, holding one of his pictures with the headline, "Artist Loses Life's Work" and he was crying. It felt as though I were looking at a newspaper that had been used as a prop in a film. I kept thinking that my kids will never see the house where I grew up. I think when sadness is at its worst you wake up and for five seconds you think everything's OK and then you have this sudden shock that something's happened. It's almost like someone's pressing an iron bar down on your chest. True sadness to me is like a physical pain. My heart actually hurt for weeks and weeks.'

It is true that sadness has physical effects on the body. Skin conductance, associated with sweating, increases and even intestinal processes can change. A patient called Tom, from a case study in 1943, had a stomach which became pale whenever he was depressed and, back in the 1920s, it was found during other research that depressed patients secreted less gastric acid into their stomachs than usual. There are also hormonal differences between the depressed and the non-depressed. About half of those with depression seem to have abnormally high blood levels of the hormone cortisol which is released during times of stress under the control of a brain system called the HPA axis. This causes a cascade of chemical reactions. A small area of the brain called the hypothalamus (H) organises activities such as sex-drive and the control of body temperature. It releases a hormone which affects the pituitary gland (P) which in turn causes the adrenal glands (A) above the kidneys, to produce the stress hormone cortisol, hence the name HPA axis. Crucially, for some people this control system doesn't seem to work, resulting in an excess of cortisol.

Naturally the main organ affected by sadness is the brain. There are changes in the levels of neurotransmitters or chemical messengers which communicate between the millions of neurons in the brain in a kind of relay race. One of the most important is serotonin which is ejected in a burst every millisecond and seems to have an involvement in almost everything that happens in the brain, without having the sole responsibility for any single function. When we feel depressed serotonin levels are lower. This is the basis on which anti-depressants like Prozac work. Prozac is an SSRI or selective serotonin reuptake inhibitor. Normally when serotonin is squirted out of the nerve ending it acts on receptors in the brain and if too much is released it's taken back up into the nerve endings so that it can be recycled. The idea of drugs like SSRIs is to block this process, so that instead of being taken back up, the serotonin can spend longer in contact with the receptor and have more of an effect on mood. The effect, however, is far from immediate. When Julia took Prozac when she was depressed it was a few weeks before she noticed a difference. 'You take it for a few days and you wonder why you're bothering because you still feel terrible. Then very gradually it starts to work. It's not as though you suddenly feel high or happy; it's more as though your real self has been covered up and something in the tablets uncovers it and lets you be yourself again.'

If SSRIs can increase the amount of serotonin which is active in the brain and serotonin can make you feel happier, then it should follow that we could all feel more cheerful if we took an SSRI. However, in a non-depressed person these drugs have no effect on mood because of the way that serotonin operates in the brain. It's not like dopamine, the neurotransmitter which causes us to feel joyful. If you are not depressed then SSRIs make no difference to the way you feel. In fact, even lowering your levels of serotonin artificially leaves your mood constant if there is no history of depression in yourself or your family, suggesting that a vulnerability to depression is the key. If you have had depression before and your serotonin levels are lowered medically then you will feel depressed once again. For some reason studies have found that women's serotonin levels are easier to lower artificially, while men's brains appear to be better able to compensate for these chemical changes induced in a lab. Phil Cowen, a psychopharmacology professor at Oxford University who has worked on SSRIs for several years, described this phenomenon as a little like a scar. Once you've had depression the scar remains. Somehow those pathways in the brain have been disrupted. It's also complicated by a natural variation in the number of receptors for serotonin that each person has. It can only have an effect if there are receptors to receive it.

a diagnosis of depression

The distinction between depression and sadness could be classified as a differentiation created by society. Extreme, prolonged sadness can reach a point where friends and family are unable to help. At that point, as with Chloe, they entreat the person to seek professional help and the symptoms are labelled as an illness – clinical depression. This diagnosis came as a great relief to Chloe, providing her with an explanation for her feelings whilst absolving others of any blame.

Symptoms of clinical depression fall into four categories, but a person doesn't have to experience all of them for a diagnosis of depression. The most obvious are the emotional symptoms. People are often saddest and most tearful in the mornings, using words like 'blue', 'hopeless' and 'lonely' to describe their mood. It can be accompanied by anxiety and a lack of pleasure in normally enjoyable activities such as eating tasty food or seeing friends. Then there are the physical changes – waking very early in the morning, loss of appetite and loss of interest in sex. The third type of symptom concerns motivation. Depressed people usually find it very hard to get started on projects, or to make decisions.

Finally and most intriguingly there are the cognitive symptoms – those associated with thoughts. Even feeling slightly sad can affect the way a person thinks. This can be illustrated by a simple word experiment: participants listen to either happy or sad music through headphones. Afterwards they sit in front of a computer and letters are flashed up on the screen. This is a standard psychological test called a lexical decision task, where the letters on the screen sometimes form a word and sometimes don't. The task is to identify the real words as quickly as possible. Those who had been listening to happy music were faster at spotting the words associated with happiness like 'delight', while the people who heard sad music identified the sad words like 'weep' the fastest. If the way the brain processes words is affected by a temporary music-induced mood then for a clinically depressed person these changes will last much longer. Unfortunately these shifts in thinking can help to maintain a person's depression.

Depressed people often perceive themselves to be useless and begin to believe that anything that goes wrong must be their fault. Imagine what you would think after you'd accidentally dropped a glass and broken it. If you're not depressed you might be briefly annoyed with yourself, but as you cleared up the fragments of glass, you would probably console yourself with the fact that it was an accident and that these things happen. A depressed person is more likely to blame themselves, to see it as further proof that everything always goes wrong for them and that it always will. The depressed person looks round for evidence to support their view that everything's hopeless, while dismissing any good fortune as pure chance. Addressing these sorts of thoughts provides the basis for cognitive therapy for depression.

Chloe found that her depression forced her to stop and take stock of her life. She felt that she had been so happy previously that she hadn't been sufficiently self-analytical. A tragedy in her life also provided something of a turning point. A few months after starting on Prozac she was feeling slightly better and went to Australia to visit a friend, but became very depressed once more. Then on 12th September 2001 she heard that a friend of hers was missing after the destruction of the twin towers in New York. She flew to New York to help his wife in her search, but he had been killed. For the first time in years, Chloe found that someone else needed her help and she couldn't be the sad person anymore. After five weeks in New York, helping her friend to arrange a memorial service, she realised that she was able to deal with practicalities and, despite her grief, she found that she could do something she had not done for a long time – laugh.

why do only some people become depressed?

Traditionally, an expert's answer to this question depends on the theoretical background in which they are working. One of the most plausible explanations and one that we can all relate to, is that social factors play a large part. It wouldn't be surprising for depression to be present in a tough life where lots of things go wrong. A landmark study conducted in the seventies by two British sociologists George W. Brown and Tirril Harris, found that the biggest risk factors for depression in women were caring for three children under the age of five, with no family or friends in whom to confide, and having lost a mother before the age of eleven.

However, your chances of getting depressed might have been set in stone long before adulthood, even before birth. If one of your parents or siblings suffers from clinical depression it makes the likelihood of you becoming depressed between one-and-a-half and three times more likely. This doesn't prove that depression is simply a genetic condition. Instead you might learn from a parent or sibling that depression is the best way to respond when things are going badly. A different person might turn to violence or alcohol abuse instead. Twin studies are the standard way of separating the influence of genetics from the effect of life experiences – in other words nature and nurture. The theory is that if identical twins are brought up separately then any differences between them must be down to the environment rather than genetics, because both twins have the same genes. If one identical twin develops depression there's almost a fifty–fifty chance that the other twin will too, indicating a role for both genetics and life experiences. However, it's not simply the case that there is one gene which predisposes you to depression, or not as far as we know. For some reason various genes seem to be implicated.

Other researchers believe that to get to the root of depression we should study chemical changes in the brain. However, this need not rule out the influence of the environment. The work on serotonin is a good example of how all these factors can work together. Studies of monkeys in the early 1990s demonstrated that the dominant members of the group had higher serotonin levels, but that these levels weren't permanent. If a monkey was removed from the group by the experimenters its serotonin levels fell. The same thing could happen in humans. You lose your job, which affects your feelings about your status, which causes your serotonin levels to fall, with the result that you feel sad. Low serotonin levels tend to have different effects on men and women; women are more likely to become depressed while there's some evidence that low serotonin levels in men lead to aggression. In the monkey studies it was found that after the dominant monkey had been removed, the serotonin levels of whichever monkey took its place would rise. Interestingly, if a monkey's serotonin levels were artificially boosted, the monkey appeared to show an improvement in social skills and began rising through the group and, provided the dominant male was absent, could even take his place. If their levels were artificially lowered they would lose status.

Trying this out on people, Alyson Bond gave them SSRIs for four weeks and then gave them various games to play as well as asking their flatmates to report back on any changes they might have noticed while living with them. During the games the people on SSRIs made more eye contact while they were speaking and less while the other person spoke – a sign of dominance. They also became more cooperative during the game, while their flatmates reported them to be less submissive than they used to be. This suggests that high levels of serotonin increased both their dominance and made them more cooperative. Although these two results might sound contradictory, in fact this fits in with the pattern seen in monkeys. Those of high status aren't necessarily more aggressive, but they are better at getting on with other monkeys.

So it appears that success can boost serotonin levels *and* high serotonin levels can lead to success, possibly through getting on with other people better, which might seem somewhat unfair. Those who are already happy and have high serotonin levels are likely to continue to succeed in society and remain happy, while those with lower levels – the very people for whom success could make a big difference – get left behind and their levels remain low. Most of the work in this area involves animals and so it can't be guaranteed that the same effects would be seen in humans. However, a study conducted in Pittsburgh in 2000 found that the people with lower socio-economic status have a blunted response to a particular drug, which in turn suggests that they have a low turnover of serotonin. Since people with lower socio-economic status are known on average to experience a greater number of stressful life events and to be exposed to more episodes of physical and psychological violence, the researchers speculate that these negative experiences could be altering the brain in such a way that the turnover of serotonin is reduced long-term.

This illustrates that the search for chemical explanations does not rule out the role that life events can play in the way we feel. Our mood is not predicted solely by the quantities of certain chemicals in our brain. Changes in these chemicals might simply be a reflection of influences from outside, so when things go wrong in life the balance of chemicals can shift. Early experiences could even play a part by affecting the way in which the brain copes with changes in neurotransmitter levels.

Earlier in this chapter the stress hormone cortisol was mentioned. If it is the case that some depressed people have a problem with the system regulating cortisol, then as with serotonin function the same question remains: why has the system gone wrong in those particular people? They might have been born with a predisposition to release excess cortisol, or once again their experiences might actually alter the system. Charles Nemeroff from Emory University, Atlanta, has found that newborn rats who were separated from their mothers for ten out of the first twenty-one days of life, grew up with increased levels of the hormone. This suggests that early negative experiences might rewire the brain in terms of its response to stress. This hypothesis is based on animal studies, so is not entirely conclusive, but Nemeroff might have finally hit on a biochemical explanation for how bad childhood experiences could link to depression in adulthood.

Not everyone believes that cortisol or serotonin holds the chemical answers to depression. There is a new theory, admittedly very much in its infancy, that the immune system could be involved. The idea is that people with problems with the immune system respond in one of two ways – either they develop an auto-immune disease or they become depressed. Women are more prone to both of these conditions. It is even possible that drugs like SSRIs are acting not only on serotonin but on the immune system and the cortisol system in addition, which could explain why the drugs take several weeks to have an effect.

Again, these chemical changes in the brain could sometimes be a response to the outside world, rather than the brain spontaneously malfunctioning by itself. It's arguable that we should no longer consider the physiological and the social to be separate. They affect each other, which would explain why drug treatments and talking therapies – which couldn't be more different from each other – can both relieve depression.

the mystery of tears

Two years before her parents' house burned down Julia was on holiday in Brisbane with her mother, who was recovering from cancer. Julia had recently called off her wedding and still felt desperately sad, but not wanting to add to her mother's troubles she put on a brave face and tried to enjoy the holiday. Night-time provided the chance to cry. Once her mother was asleep she lay in the next bed weeping silently. Two weeks into the holiday her mother asked her why she was crying so often, admitting that she had listened to her sobbing every single night and didn't want to be protected from Julia's sadness just because she was ill.

Julia was grieving for the loss of the future she'd imagined for herself and was keen not to add to another person's distress by discussing it. These responses both make sense. What is stranger is that a clear liquid should fall out of her eyes because she's unhappy. It's easy to see how tears wash out your eye and protect the surface from a sharp speck of dust or an eyelash, but why do we cry when we're upset? In physical terms a bout of sobbing blocks the nose, irritates the eyes, puffs up the face and makes the head ache, yet sometimes we can't help but weep.

A group of volunteers file into a makeshift cinema in a lecture theatre at the St Paul-Ramsey Medical Center in Minnesota, USA. They sit and watch a film about some children who are caring for their dying mother, only a year after their father died. It's a film which openly manipulates the emotions. The phrase 'tugging at the heart strings' could have been invented for it. After they've lost both parents the children stand in a line in the snow while the eldest boy bravely declares that he will look after his young brothers and sisters. Meanwhile the heartless elders of the town just want rid of the lot of them. The volunteers watching the film have never met. They sit in silence wearing special goggles which have miniature buckets suspended beneath the eye-pieces. The reason they are here is to provide tears.

After years spent studying crying Professor William Frey has found that this is the easiest way to make tears come forth. Unfortunately the special glasses didn't work because tears escaped down the criers' cheeks. In the end he found it was easiest to have people collect their own tears in tiny test tubes. He experimented with different films and seating arrangements, eventually concluding that spacing was the key. When people are close to a stranger they hold back the tears.

Professor Frey went to all this trouble because he wanted to know whether tears of sadness contained different chemicals from the tears we cry when our eyes water due to soreness or irritation (known as irritant tears). He was drawn to the topic because he himself hadn't cried since he was a boy and felt he might be missing out. Studies have shown that, like Professor Frey, some people don't cry at all in a month, while it's been found that others cry on up to twenty-nine days of the month. There are of course social rules prescribing when it is and isn't appropriate to cry, particularly when it comes to boys. When my father was in pain in hospital after a tonsillectomy at the age of four, a nurse told him that boys don't cry and that if he did people would think he was a girl. In fact despite this kind of social pressure, boys cry just as often as girls until the age of twelve, but by adulthood women are crying four times as often as men. Teenage girls are no more depressed than boys which has led to suggestions that the sudden difference in their crying frequency from the age of thirteen might be caused by the increase in girls in the hormones oestrogen and prolactin. However, studies on crying during pregnancy and pre-menstrual crying have been inconclusive, making hormonal explanations tricky. Of course it's been assumed that the teenage boys are behaving normally by crying rarely while the behaviour of teenage girls needs some explanation, hence the hormonal theories. If you turn this idea on its head, perhaps there's something about puberty that stops boys crying. Is it possible that increased levels of testosterone somehow block the crying response? When small mammals were given testosterone injections they vocalised less. It is also the case that as men get older they begin to cry often once again, just as their testosterone levels drop. Alternatively maybe they start holding back the tears when they reach adolescence for fear of someone seeing them and by the time they reach old age they simply worry less about a witness to their crying.

Weeping rates also vary from culture to culture. A researcher, Marleen Becht, spent three years collecting data from twenty-nine countries. She found that both men and women from the United States cried the most often, while Bulgarian men and Icelandic and Romanian women cried the least. It's hard to know exactly what to make of that. For a start in some countries only thirty individuals were asked, which can't tell you much about the habits of an entire nation. Moreover, these comparisons are based on each person's own estimate of the number of times they had cried in the last four weeks. Cultural attitudes towards crying might influence the amount of crying to which people are prepared to confess. Having said this, when I visited Iceland I couldn't help thinking about the rarity of tears there and found myself watching to see if I could spot any moist eyes. Disappointingly the guide didn't burst into tears at the sight of a beautiful, white, double waterfall, but to be fair, he probably visited it every day.

Research also tells us that the most likely time for crying is between 7pm and 9pm. This isn't surprising; it might be the first time all day that a person has had any privacy and the factors likely to induce crying are all present – tiredness, sad TV programmes and family arguments. There's also the possibility that the circadian rhythms which govern our sleep cycle play a part. Babies cry the most in the evenings, why not adults?

Tears of sorrow seem to be unique to humans, although there are anecdotal reports from Darwin of Indian elephants weeping when they are bound and immobilised. Professor Frey wrote to lots of zoologists and animal trainers to see whether they had witnessed such events. Most said that they had not but he has received anecdotal reports from pet owners who report seeing all sorts of animals shedding tears – from pigs to Chihuahuas. Diane Fossey, who famously spent years studying mountain gorillas, described witnessing Coco, a three-year-old captive gorilla, looking out of the window and shedding actual tears. I was tempted to write 'looking out of the window longingly', but that would have been my interpretation of the gorilla's emotions and therein lies the problem. The trouble with these reports of weeping animals is the all too human tendency to anthropomorphise. It's hard to distinguish what an animal is feeling from the way we imagine we would feel in the same situation.

zeis, manz and the crypts of henle

In humans, at least, we do understand something of the chemical constituents of tears. They were first analysed back in 1791 and were found to consist of salt, mucus and water. Most of the watery part comes from the lachrymal glands which are at the top outer edge of each eye. This liquid is sandwiched between inner and outer layers which come from glands with such wonderful names as Zeis, Moll, Manz and the crypts of Henle. The oily top layer stops the liquid from evaporating too fast. When we blink tears slide across the surface of the eye towards the inner corner, but if we cry the tears collect in triangular lakes until they overflow, slipping over the edge of the eyelid and down the cheek.

During the first two or three months of life newborn babies don't shed tears when they cry. Darwin believed that as babies grow older they develop the ability to cry in order to protect the eyeballs when they screw up their faces in distress. He noted that in every culture extreme laughter, yawning and vomiting also brought tears rolling down the cheeks and concluded that the cause must be the same. When the face is screwed up, he reasoned, the blood vessels around the eye become engorged and tears are needed to protect the surface of the eye. He decided to capitalise on his children's propensity for tears to test his theory, asking them to contract the muscles around their eyes as tightly as possible for as long as they could, believing that this would induce the production of tears. However, none were forthcoming. He did not let this dissuade him from his theory, concluding that his children were simply unable to produce voluntary contractions of sufficient strength. Tears of sorrow must exist to protect the eyes, he contended, because identical tears are shed when a speck of dust is lodged in the eye.

This is where Darwin was wrong and his assumption takes us to the heart of William Frey's research. As well as devising methods of inducing sadness to make people cry, Frey experimented with substances which would irritate the eye sufficiently to produce tears. Ammonia and tear gas were ruled out for ethical reasons. Instead he gave his volunteers various other substances to inhale including fresh horse-radish, but in the end it was the old cliché – onions – which really made them weep. Even in Shakespeare's time these were the old crying standby. In *The Taming of the Shrew* a boy who is acting the part of the woman is advised,

> And if the boy not have a woman's gift,
> To rain a shower of commanded tears,
> An onion will do well for such a shift.

As well as watching over-sentimentalised films, Professor Frey's brave volunteers had to put their faces over a blender full of freshly chopped onions and inhale deeply with their eyes open for about three minutes. The moment an onion is cut a substance called thiopropanal-S-oxide is released into the air. When it reaches the tear film on the eye there's a chemical reaction which produces sulphuric acid, so not surprisingly it stings. The only way to stop the reaction is to cover up the eyeball or to stop the substance escaping into the air in the first place. Hence the two successful methods for avoiding streaming eyes are washing the chemical away by peeling onions under water or wearing contact lenses to prevent the vapour reaching the surface of the eye.

When Frey came to analyse the two types of tears he found that tears are not all the same; the emotional tears contained 24% more protein than the irritant tears. The purpose of these proteins in the tears is to fight infection and to control the levels of acidity on the eyeball. This suggests that something special is happening when we shed emotional tears – an expulsion of chemicals perhaps. For this to have any emotional benefit these substances would need to have an association with stress, but as far as we know there isn't a clear link between stress and these proteins. However, when Frey dissected whole tear glands he did discover the presence of two hormones known to be related to stress – ACTH and leucine-enkephalin. The former is also found in tears themselves. He believes that when we cry we expel toxic substances which are by-products of the stress we're experiencing. The idea is that you flush them out through your eyes with the result that you feel slightly better. Indeed he did find that 85% of females and 73% of men reported feeling better after a good cry. Even the existence of the phrase 'a good cry' suggests that it's seen as useful. I saw a TV advert recently for a CD of '*All-Time Classic Tearjerkers* – the most moving tunes for times of reflection'. The very title accepts that sometimes people want to cry and these sad tunes might help them along.

However, attempts to demonstrate the beneficial effects of crying in a laboratory haven't quite worked. It ought to be simple. Ask people to rate their emotional state, show them a sad film, wait for them to finish crying and then ask them to rate their emotional state again to see whether there's been an improvement. Unfortunately there's usually no difference, suggesting that it's not the expulsion of toxins through tears that makes you feel better. Perhaps you only feel better if those tears encourage someone else to comfort you. There is a big problem, however, with extrapolation from such an artificial situation to real life. When you cry during a sad film you are simply empathising with the characters and imagining yourself in that situation, which is rather different from feeling so helpless in your own life that you cry. Moreover, the situation is inevitably going to affect the way you feel after crying. At home a good cry might lead to a sense of relief, but at work or in a laboratory you might end up feeling awkward or embarrassed.

There is another possibility for the discrepancy between people's own reports of feeling better after crying and the laboratory experiments. The mind doesn't assign equal importance to the creation and storage of memories of different events. We all have biases that affect what we remember. It helps us to justify our crying behaviour if we only recall the times it made us feel better.

the message of tears

Randy Cornelius sat in a studio at his local radio station WSPK in Poughkeepsie, USA – the radio station promising to play 'today's best music'. He's a psychology professor at Vassar College and was waiting to be connected to the gloomy cupboard of a studio where I was sitting at the BBC in London, waiting to interview him for a radio series. It took a while and I could hear various engineers from both the British and American ends come over the headphones. Then, another voice with an American accent, 'Hello, hello, can anybody hear me?' I wasn't sure whether I'd reached the man himself or another engineer. 'Are you Randy?' I asked. 'I sure am,' he said, causing mirth at the British end, but not a flicker from the States, where it doesn't have quite the same meaning. Once it was established that the link between the studios was working, we went on to have a serious discussion about crying.

In contrast to Professor Frey, he believes that tears are all about communication; they let other people know that you're upset, information which might ultimately benefit you. Crying could be a powerful way of telling another person that their 'harmless' teasing has in fact touched a nerve and that they should stop. It also signals to those around you that you need their sympathy or help.

After Julia and her fiancé split up she was walking along the street sobbing uncontrollably when a stranger approached her offering help. Because she was crying so much she couldn't answer and shook her head. Tears provide such a strong message that they can even elicit help from strangers. Cornelius believes the failure of the laboratory studies to demonstrate benefits from crying is due to the fact that during the experiments nobody receives any comfort from another, so they haven't gained any help by crying and therefore don't feel any happier afterwards.

Think of the kinds of situation where you tend to cry. When Randy Cornelius asked people to do this the occasions most often cited were the death of a friend, the end of a relationship, watching a sad film or poignantly happy events like a wedding. Now think back to the last time you actually did cry. Here it was a slightly different story. Tears tended to follow arguments or rejection or feelings of loneliness or inadequacy. Cornelius believes helplessness might be the key reason we cry; we feel we can't do anything to change the situation, so we cry. Or to put it another way, we can't do anything more for ourselves, so we need other people's help and it's crying which signals the seriousness of our situation. Babies cry in order to get the attention and help they need. Perhaps adults are doing the same.

the quintet of the astonished

In the second exhibition room at the National Gallery in London there are three cool cream benches, lined up one behind the other. We sit and watch in silence, apart from the inevitable occasional cough. On the wall in front of us there's a life-size photograph of five people standing in two rows – a man and a woman at the front, and three men behind them. Although they are close together they don't look at each other. Three of the people are staring at the same place in the middle distance. The woman has her arms crossed in front of her chest, one hand on top of the other. Slowly her left fist clenches tightly. This isn't a photograph after all. The people are moving, very slowly. The woman looks angry, yet despairing. Who is she gazing at? She opens her mouth and her shoulders rise. She is trying to control her emotions but still looks tortured with distress. What terrible thing could she be watching? Has she just seen the man who killed her child? Behind her shoulder, a man has his eyes closed while he smiles in beatific joy. His face is full of contentment, a sublime, holy contentment. The man at the front screws up his face in misery. Is he calling out? His hand moves up in front of his chest. He is distraught. Is he the husband of the distressed woman? He doesn't even glance at her; he seems to be in his own world of agony. Another man has his hand on the woman's shoulder, but despite her distress he is almost smiling. He looks proudly happy. He could be watching his child playing a musical instrument. How could he be so happy when two of the others are so distressed? The woman's bottom lip starts to jut out in anger. Her hands are clasped together so tightly that the veins on her forearms stand out making dark shadows down her wrists. The tortured man continues to cry out. The man next to him puts his hand on his shoulder but takes no notice of his distress, continuing to stare into the distance. The moving photograph is in a frame and is

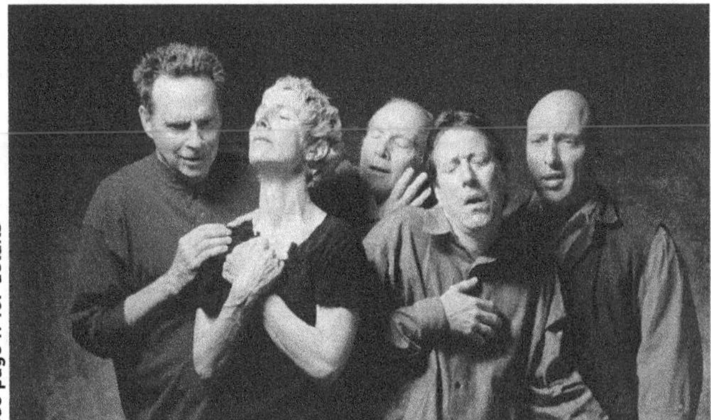

see page iv for details

lit like an old master. Light falls on the faces and there are dark shadows in the folds of their clothing. However, this isn't an old master. It's an artwork by Bill Viola called *The Quintet of the Astonished.*

It takes fifteen minutes for the emotions to play themselves out on the screen in slow motion. It's uncomfortable to watch, but not because two of the people are so distressed. We know they are actors playing a part for the artist to film. The reason the film is discomforting is that you yearn for them to comfort each other. The man in religious ecstasy is so close behind the woman that at some points it's almost as though he's inhaling the smell of her hair and smiling in appreciation. They stand so close together that they are touching, but they never interact or even make eye contact. The man who is enjoying himself turns towards the distraught woman and puts his hand on her shoulder, but while you long for him to look into her eyes and show her that he's there for her, he continues to smile instead. You find yourself wishing that if no one else is going to help the two tortured people, they could at least turn to each other and suffer together, but they are destined to suffer alone. The fact that it is so painful to watch these people standing so close and yet ignoring each other's emotions shows us something fundamental about emotions – that they are an exceptionally strong form of communication. Crying in the presence of another is so powerful that it is unbearable to watch if that display of emotion is ignored.

If the purpose of tears is to communicate your sadness so that others will help, there is just one problem: people often cry on their own and report feeling better afterwards even though they have received no comfort from others. Moreover people often deliberately seek privacy if they want to cry. It has been suggested that even if you cry alone, you're using yourself as an audience. You might give yourself comfort by sympathising with yourself, agreeing that you have the right to feel unhappy about your situation in the same way that a friend would.

It is possible that the opposing theories of Professor Frey and Randy Cornelius are in fact compatible. Perhaps we gain some relief from expelling the toxic by-products of stress *in addition to* communicating our distress and receiving comfort from others.

If crying can be beneficial, this raises the question of whether never crying, like the Professor of Tears himself, could be harmful. In the Western world there has long been an idea that if you suppress your tears you will do yourself physical damage. Back in 1847 Alfred Lord Tennyson wrote in the poem 'The Princess',

> Home they brought the warrior dead;
> She nor swooned nor uttered cry.
> All her maids, watching said,
> 'She must weep or she will die.'

It has been suggested that suppressing the emotions could increase a person's chances of developing cancer, heart disease or high blood pressure. Related studies tend to focus on emotions in general and often anger in particular, rather than crying. However, in one study people were instructed to watch films while suppressing their tears and laughter.

This was found to increase heart rate, so in theory if this behaviour pattern was repeated over many years a person's health could be affected. Nevertheless, you are unlikely to be in situations where you are suppressing tears many times every day, so not crying probably only makes a difference to your health if you are suppressing all your emotions all of the time.

In eighteenth-century France collective tears were seen as enjoyable and compassion was expected from others, but by the nineteenth century suppressing tears was seen as evidence of self-control. A social dictionary of the time describes tears as 'water too often ill-employed, for it remedies nothing'. When attitudes towards crying have been researched in more recent times, findings using questionnaires contrast sharply with the results of experiments held in the laboratory. In questionnaires women tend to say that they feel sympathetic when someone else cries, whereas men report feeling awkward or even manipulated. Women in particular seem to be accused of 'turning on the waterworks'. Men report believing that it is inappropriate or a sign of weakness for a man to cry, but in laboratory experiments where people are actually crying the results are rather different. When a man cries after watching a sad film, both men and women consider him to be more likeable, while the weeping women are disliked for their tears. So although men may not like the idea of other men crying, when actually faced with it, they do not judge them. Perhaps men consider that if a man is crying his feelings must be profound and so he should taken seriously.

If crying succeeds in communicating our feelings then

once again sadness is revealing that it does have a purpose. The family seeing off the woman with the linen jacket in my airport story, eyes wet with tears, would be signalling their sorrow. While that parting would be reflected in their brain chemistry, they would in fact be strengthening their bonds with the woman who was leaving. She would be able to see how much they cared, in a way they probably wouldn't usually show her.

Although we don't tend to think of it as a pleasant emotion, sadness is probably one that we are stuck with and although the value of sadness is hard to see, particularly when we are feeling sad, it might be an emotion with wisdom after all; an emotion which forces us to slow down, consider our plans and maybe change them. Sadness provides a light and shade in our emotional life. At the same time the outward signs of sadness like a down-turned mouth and that most potent but still mysterious communicator of sadness, tears, can bring us closer to those around us by signalling to them that we need them.

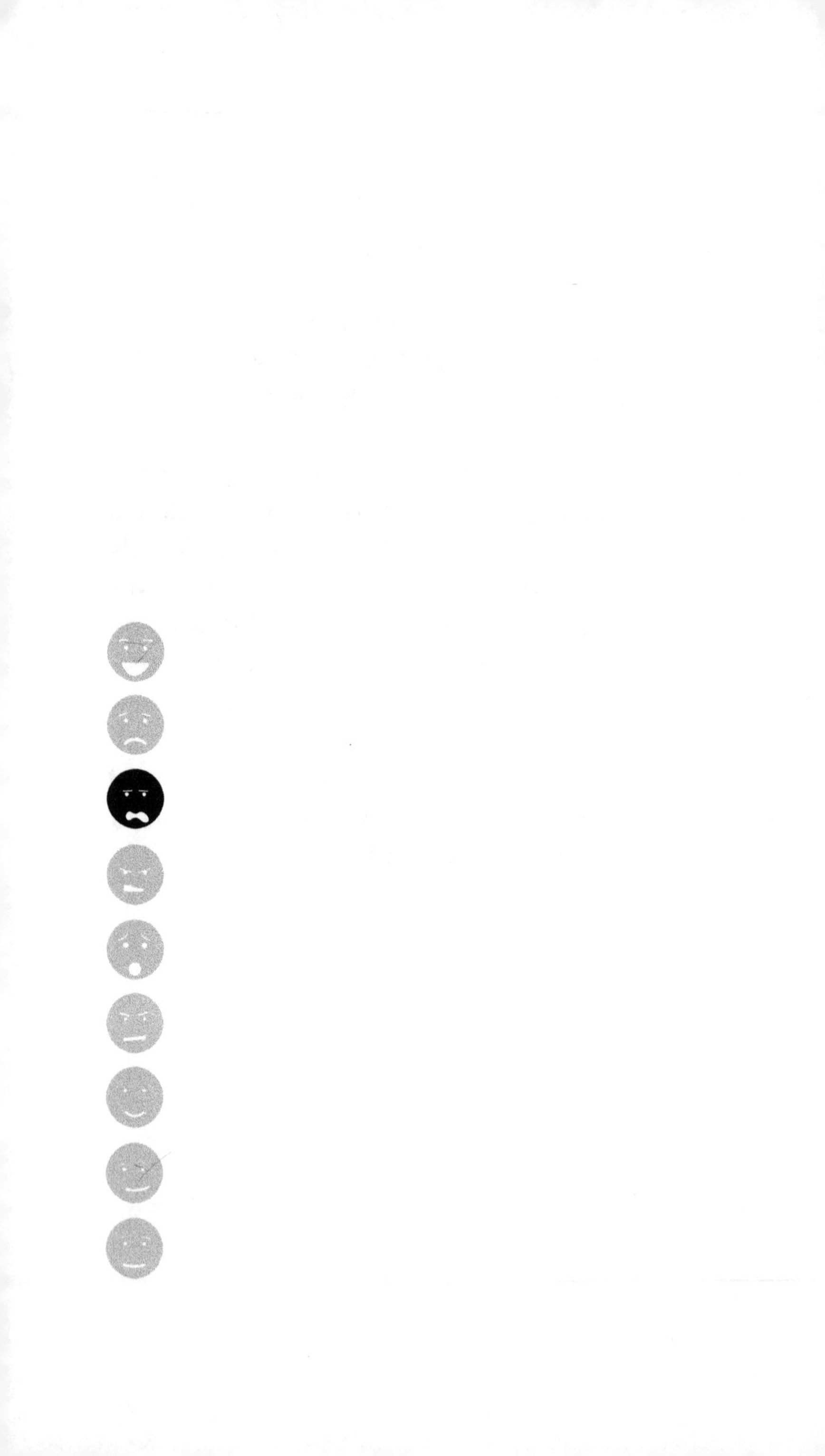

three

disgust

The audience sits in silence in a converted warehouse in East London. Although it's a cold November night, waiters further up Brick Lane stand outside on the street, trying to persuade passers-by to choose their particular curry house. But in the warehouse eating is the last thing anyone would want to do. Out of 2,000 people who tried, these are the lucky 300 who succeeded in getting tickets. A few watch anxiously as a man comes onto the stage wearing the curious combination of a black Fedora hat and surgeon's scrubs. His name is Professor Gunther von Hagens. After a short introduction his assistants wheel in a long, sheet-covered lump on a trolley. The professor draws back the white cover to reveal the dead body of a man with skin which looks as though it could be made from plastic. The professor walks over to a side-table on which is a silver tray lined up with implements, ranging in size from the smallest knife to a hacksaw. He selects a scalpel, turns back to the body, leans over the man's breastbone and puts the blade into contact with the skin. He presses down firmly and slices down through the skin. The audience wince as one, imagining the knife cutting through their own flesh. Surprisingly there's no blood, just a slow trickle of thick orange liquid, like the orange congealed fat left behind in a roasting tin. This is the first public autopsy to be held in Britain for 170 years and in addition to the live audience, millions are watching on television. Professor von Hagens explains that he's making what's known as a Y-cut, slicing across the chest and down the centre of the torso. The man died at the age of seventy-two after drinking two bottles of whisky a day and smoking heavily for years. As the chest is opened the skin is peeled back on either side of the cut to reveal layers of fat. The heart and lungs are extracted and carefully placed in silver dishes lined up on the side-table. Like waiters at a banquet, assistants stand in a line nearby, ready to pass a dish when required for the next body part. Eventually just one dish remains empty. It's time for the brain.

An assistant holds the man's head still, while von Hagens carefully cuts around the head from ear to ear, loosens the skin enough to slide his hands in behind it and peels back the skin to expose the skull. Taking a hacksaw, he begins grinding his way into the skull, explaining to the audience, as he cuts, that due to the skull's three layers, this can take some time. When he hears a change in tone he knows he's through. 'I am about to take the brain out,' he calmly announces, as though it's a cookery demonstration. The brain comes away surprisingly easily. He simply picks it up and lifts it out without resistance, like a walnut out of a shell. Nobody in the audience speaks. Their brows furrow

and they lift their hands up to their faces, covering their mouths and half-masking their eyes. They are experiencing disgust.

Although it is such a basic emotion, disgust is often forgotten; if people are asked to list some common emotions it's usually a long time before disgust is suggested. Yet of all the psychologists I've met who research different emotions, those who study disgust seem to do so with a particular passion. They told me that they have learned one thing, however, and the same applies here: however fascinating disgust might be, if you want to enjoy your food it's not a good idea to read about disgust while you are eating.

There is one way in which disgust differs from many other emotions: feelings of disgust always have a clear cause. You can't wake up one day feeling generally disgusted in the same way that you might feel generally sad. There has to be an object of your disgust and, as we'll see, it's these objects which provide clues as to the purpose of this strange emotion.

The loos at the Glastonbury Festival are infamous. By the end of the weekend after thousands of people have used them, they get very full. The story goes that every year, on the last day of the festival the same trick is played on one very unlucky toilet user. A group of people wait until a man has locked himself into the cubicle and then they tip over the entire box so that the unfortunate inhabitant is trapped lying in the contents of the now emptied toilet.

Just hearing about this story may well provoke a physical response in you. It certainly would for the victim of the trick. Disgust is a particularly visceral emotion. It can make you shudder, salivate, feel physically sick, retch and, at its most extreme, vomit. The facial expression for disgust is particularly distinctive: the nostrils narrow, the upper lip rises high, the lower lip lifts and protrudes slightly, the cheeks rise, the brows lower creating crow's feet beside the eyes and the sides of the nostrils ascend, causing the sides of the nose to wrinkle. When a person is disgusted other people can tell exactly what they are feeling from those sneering lips. This is the disgust face and it appears very early in life.

how we learn to feel disgusted

Disgust is one of the earliest emotions that we experience. From birth, babies show disgust at bitter tastes and as Charles Darwin noted the expression of disgust gradually becomes more frequent. He became fascinated with the development of emotions after the birth of his first child and decided to document his son's emotional expressions. More than thirty years later he wrote a book on the subject, a book which is often overlooked today. Darwin clearly observed disgust on his son's face at the age of five months – on one occasion in response to cold water, on another at a piece of ripe cherry. 'This was shown by the lips and whole mouth assuming a shape which allowed the contents to run or fall quickly out; the tongue being likewise protruded. These movements were accompanied by a little shudder. It was all the more comical, as I doubt whether the child felt real disgust – the eyes and forehead expressing much surprise and consideration.'

Whether or not babies do feel the emotion in the same way as adults, by the age of three toddlers have learnt what adults consider to be disgusting. Although babies and toddlers will happily play with their faeces they soon discover that they shouldn't, suggesting that they only learn disgust through social conditioning. This is reinforced by a study which examined fifty children who had grown up in the wild; none showed any signs of disgust at bodily products. However, toddlers tend to learn disgust at faeces particularly easily, leading some such as Val Curtis, who researches hygiene and disgust at the London School of

Hygiene and Tropical Medicine, to suggest that we are born with a propensity to find certain objects disgusting; it's far easier to convince children to be revolted by faeces than by sweets. Having said that, it is impossible to disentangle the influence of socialisation. Inevitably it would be difficult to train a child to believe that sweets are disgusting because other children would soon undermine this view, whereas disgust at faeces is reinforced by every adult or older child they meet.

As children develop, their responses to disgust become more sophisticated. Imagine you go into a room where you're given a glass of clean, fresh water. You drink some water and then you're asked to spit into the glass before taking another sip. Would you do it? It is your own saliva after all and only moments before it was in your mouth. Then you're given a fresh glass of water and a dead, but sterilised, cockroach is held in tweezers and dipped into the water. There's nothing physically wrong with the drink in either case but when university researchers gave people these tasks most wouldn't do it. At the age of four children will happily drink up, but by seven, like adults, they don't want to. Before this age they might not possess the complex thought processes which would allow them to see contamination in the same (admittedly at times irrational) way as adults. An understanding of contamination requires the ability to follow a long chain of events. In order to feel repulsed by the idea of licking an object which has fallen on the floor, you need to consider that somebody who had previously stepped in dog faeces would have a dirty shoe which then touched the floor, thus contaminating the food, and finally you. By the age of seven or eight children's thinking skills have developed to an extent where they can not only follow this chain, but can use disgust to their advantage with ploys such as licking the last biscuit and then offering it to their squirming sibling with the words, 'Go on – eat it then!'

the disgusted brain

The fact that the facial expression for disgust is so striking reflects the significance of our ability both to convey disgust and to detect it in others. If one person tastes contaminated food everyone else needs to know to stay away from it. The disgust face is so central to this communication that the brain has a specific mechanism for detecting disgust in others. Mary Phillips and her colleagues at the Institute of Psychiatry in London asked volunteers to lie in a brain scanner while they were shown photographs of people displaying facial expressions of either disgust or fear. Cleverly, Dr Phillips set them the task of deciding whether the photograph was of a man or a woman in order to distract them from focusing on trying to identify the expression. Despite the fact that the emotion expressed was irrelevant to the task, the scan nevertheless demonstrated activity in different parts of the volunteer's brain depending on whether the person in the photograph looked frightened or disgusted.

Deep inside the brain there's a walnut-shaped area called the amygdala which has tended to be thought of as the seat of all the emotions but with disgust this proved not to be the case. Instead, two areas of the brain are stimulated by disgust – the *basal ganglia* and the *anterior insula*, which are both very old parts of the brain in evolutionary terms. People with Huntingdon's disease have difficulty recognising expressions of disgust which is logical since Huntingdon's damages the *basal ganglia.* Extraordinarily, even carriers of the Huntingdon's gene who do not yet have symptoms of the disease have a reduced capacity for spotting the expression of disgust.

If you were to take a brain and peel back the temporal lobe or side of the brain and look deep inside, behind the ear, you would find a large pyramid-shaped structure known as the *insula*, a name derived from the Latin for island. The front of this pyramid or *anterior insula* is the area which responds when we taste strong flavours like salt. This neural link between disgust and taste is intriguing because it lends weight to the idea that disgust exists to protect us from contaminated food. The fact that disgust is found in such an old part of the brain might explain why it is an emotion which is so hard to overcome even when you know there's no reason to find something disgusting. Even infants who have been born without functioning cerebral hemispheres show expressions of disgust at bitter tastes. This type of instinctive disgust involves no thinking, reflected in the fact that the *insula* is not associated with brain regions involved in thought and reasoning. This suggests that our brains are hard-wired for us to learn about disgust.

individual responses to disgust

Although we all experience disgust at some time, the strength of the feeling can vary considerably. When Paul Rozin, a world authority on this emotion, measures sensitivity to disgust he finds at one end of the scale people who would be happy to eat live locusts, while at the other end there are people who are not prepared to blow their nose on a brand new piece of toilet roll due to its association with dirty toilets. Women tend to be at least 10–20% more sensitive to disgust than men and that sensitivity also changes over the lifespan, peaking in the teens and tailing off gradually towards old age. From an evolutionary perspective it has been argued that disgust decreases with fertility because a person no longer needs to keep themselves or their offspring healthy in order to continue the species. However, it could be simpler than that; perhaps we become inured to supposedly disgusting sights through caring for dependants. Moreover, as people reach adulthood they gradually worry less about other people's perceptions, so you might expect a corresponding reduction in the fear that other people might find you or your behaviour disgusting.

Ultimately, a little variation in our individual sensitivity to disgust doesn't seem to matter and it's unclear whether the most sensitive people do succeed in avoiding disease more successfully than others. An extreme excess of feelings of disgust can result in an obsessive-compulsive disorder (OCD) involving strict washing routines and lengthy rituals. A person might wash their hands in a precise order, even cupping water in the hand and splashing it over the taps to avoid recontamination when they turn the taps off. These rituals can become so time-consuming that they are disabling. A person might spend every morning scrubbing the kitchen to ensure that it's definitely germ-free and, in time, might become unable to leave the house because nowhere else is sufficiently hygienic.

The American movie mogul Howard Hughes became so obsessed with the avoidance of germs that he employed staff whose job it was to keep him from contamination. He devised precise rules such as using at least fifteen tissues to open the cabinet where he kept his hearing aid. Before handing him a spoon his servants had to cover the handle in tissue, seal it with tape and then wrap a second tissue on top. His obsession with dirt ruled his life, eventually leading him to live as a recluse.

While researching the way the brain processes disgust, Mary Phillips wondered whether the people with this particular type of OCD might show differences in brain activity. She scanned the brains of people with and without the disorder whilst they looked at a series of pictures that most people would find disgusting, such as photos of filthy toilets and mutilated bodies. In amongst these she added some photos which only those with washing obsessions would be likely to find repulsive – a plate covered in tomato sauce, an unmade bed. The results were striking. The *insula* was activated in everyone when they saw the disgusting pictures, but it also lit up in people with OCD when they saw the harmless pictures of domestic untidiness. What we can't tell from this experiment is which came first. Do the people with OCD have an overactive *insula*, causing them to feel the same degree of revulsion on seeing a dirty plate that the rest of us might feel when we see a dead animal? Or is it the other way around? Is the person so anxious about dirt that this is reflected in their brain activity? Dr Phillips' forthcoming study might bring us closer to the answer. She's planning to study people's brains before and after treatment for OCD. Once they have recovered, the activity in the *insula* should, in theory, reduce. However, this still won't tell us why the problem started or whether the *insula* can simply go wrong by itself.

One curious finding was that people with OCD also have trouble spotting expressions of disgust in other people. This is surprising because you would expect those with a particular sensitivity to disgust to be alert to other people's warnings that there's something disgusting nearby.

An over-sensitivity to disgust might also be implicated in some phobias. When people have irrational fears an element of disgust is sometimes involved. To compare a mild phobic situation with one of more rational fear, while the thought of a large spider makes me shudder with horror, the idea of being stranded, dangling from a rope on a steep mountainside does not make me feel revolted – although it does make my heart beat faster. And intriguingly, it has been found that people with a phobia of blood respond not with a fast heart rate as they would in other situations that *scare* them, but with the slowed heartbeat you would expect with disgust. Research has found that people who are afraid of spiders do tend to be at the higher end of the disgust sensitivity scale – with those more likely to be horrified at the idea of sharing a bottle of water, for example. In one experiment Peter De Jong from Maastricht University talked people through three scenarios. In the first you had to imagine you were a care assistant whose job it was to go into a room where an elderly man was vomiting, clean him up and change his shirt. As soon as you have finished he is sick again. In a second scenario you had to picture yourself descending some stairs into a cellar to fetch a bottle. Spiders' webs hang from the ceiling and you duck as a web touches your face. You see a spider lowering itself down on a string and then feel something touch your neck. When you take a bottle out of the crate, you disturb a large spider which runs across your hand. The third situation was more pleasant; you are simply waiting to meet a friend at the station. During this guided imagery the experimenters recorded the extent to which each person moved their *levator labii superioris* – the sneering muscle which indicates disgust. It was found that the people who were phobic about spiders found the spider scenario the most disgusting. You might have expected them to be afraid, but instead their facial expressions showed disgust. Strikingly, spider phobics also scored higher on a scale measuring general sensitivity to disgust. It seems that spider phobia might actually highlight a fear of contact with something disgusting.

In contrast to an over-sensitivity to disgust, some people experience very little. People with depersonalisation disorder are unable to experience positive or negative emotions, disgust included. When their brains are scanned the *insula* fails to respond to disgusting pictures. Somehow their brain suppresses the usual responses.

the suppression of disgust

With effort, unless suffering from OCD, we can all suppress feelings of disgust if we really need to. Parents soon become used to changing their baby's nappies and will even discuss their baby's latest bowel movements during mealtimes, often to the horror of any non-parents present. To protect their child they also happily pick up their baby's dummy from the ground and suck it clean before handing it back, something they wouldn't do under other circumstances. We are supposedly horrified at the idea of any substance which might not be clean getting into our bodies, until we want to have sex with someone and then we suddenly seem to overcome it. The American law professor William Miller, who wrote the in-depth treatise *Anatomy of Disgust*, suggests that this might actually be a benefit of disgust; because disgust is such a strong emotion, the lifting of that disgust is a sign of intimacy which could further relationships.

Sometimes the ability to ignore disgusting surroundings is simply a necessity. I remember visiting the first class waiting room at New Delhi railway station, a huge high-ceilinged room with carved wooden benches and mirrors along the length of the walls. Women dressed in beautiful saris were brushing each other's hair, exquisitely turned out, their forearms obscured by rows of gold bangles. Assuming that all these smart people would demand clean toilets I became hopeful about the state of the loos, but I was wrong. They were of the iron-step variety and there was wet faeces smeared across both the white-tiled walls and the floor. I had to take great care to keep my shorts and T-shirt clean and could not imagine how the women in long saris managed to emerge from the cubicle looking so tidy. As well as remaining immaculately dressed, their faces seemed unruffled, with no sign of disgust. They no doubt lived in spotless homes, but through necessity they had learned to suppress their disgust in public places.

Research has found that people working in healthcare have the lowest levels of disgust sensitivity. Over time they grow accustomed to revolting sights and smells and of course they are a self-selecting group; anyone who knew themselves to be exceptionally squeamish would presumably have chosen a different field.

the purpose of disgust

It was our first date. We travelled to London, planning to shop in Covent Garden, see a film and then have dinner. Everything was going well until we got to the film, Peter Greenaway's *The Cook, the Thief, His Wife and Her Lover.* Unfortunately most of the film centres on lavish dinners, animal carcasses and brutal violence; a man murders a keen reader by stuffing his mouth and nose with pages from his book until he suffocates; a naked couple are forced to hide in a butcher's van filled with dead pigs and cows. As the van turns corners the animals swing on their hooks, covering the couple's naked skin in blood and dark faeces. After watching this film there was no way we could eat anything, especially not meat. Our overwhelming feeling was disgust – disgust at anything to do with food or human flesh. Four hours later the most we could stomach was a bag of chips, which we ate in the car on an industrial estate on the way home – not an ideal first date.

In situations like this disgust seems a rather strange emotion. We turned down perfectly hygienic food for no practical reason, something we in fact do every day although we don't realise it. In every country people eat only a tiny percentage of the edible foods available. In Britain, for example, we could eat squirrels, bats, beetles, dogs, cats, mice and slugs, but we don't. In fact the number of edible creatures that we refuse to eat far exceeds the number we do eat. Whole categories of animals are not considered appropriate to eat – those kept as pets, feared animals like spiders and snakes, and any creatures such as rats who feed on rotting flesh or human waste.

When I went on a first aid course we were given dummies in shades of plastic ranging from pale pink to dark brown. They reassured us that the dummies had all been disinfected at the end of the previous day's course. After we watched a video on artificial resuscitation it was time to try it for ourselves. 'Off you go then. Check for danger. Are they responsive? Listen for ten seconds – are they breathing? That's right, call the ambulance. Now give them two breaths. Put your mouth right over theirs and get a good seal.' Having obeyed all the other instructions on cue, at this point everyone paused, leaving their dummy to die. No one quite wanted to put their mouth on the plastic dummy where they knew a stranger's lips had been just the day before. 'Come on, you're losing vital time. They'll be dead by now!' In the end we all did it because we knew we had to practise, but to me that instinctive pause was telling. There's clearly a strong psychological component to disgust. Even if you know the object is safe, as with the sterilised cockroach in the glass of water, the instantaneous feeling of disgust is hard to overcome. The psychological element of disgust is so strong we can even imagine disgust on behalf of someone else. Take the disgruntled waiter who spits in a customer's soup. The customer, unaware of the waiter's unpleasant addition, feels no disgust, but for the waiter disgust is so psychological that just knowing what he's done satisfies his apparent need for retribution. Often it is better not to know. I remember when I worked as a waitress in a hotel, I brought some used plates back into the kitchen on my first day and began scraping the leftovers into the bin. The chef soon stopped me and instructed me to retrieve the fennel from the bin for use on the next person's plate. Likewise leftover potatoes were picked off used dishes to be passed on to the next customer and, judging by the fact that nobody working in the kitchen was surprised, I had to assume this was usual.

the bag of disgust

Outside it is thirty degrees. Inside it is even hotter; I'd asked to turn off the air conditioning and shut the windows because I was recording a radio interview and the machine hum and traffic were too loud. Val Curtis has a narrow office with bookshelves along one side. As I glance up I see books on hygiene, epidemiology and infectious diseases and then something rather different – a life-size plastic model of some human faeces, sitting in a squat coil, twisting up to a point. Cheerfully Val tells me the reason for the coil-shape is that it was bought in Holland where loos have shelves in the bowl, so rather than dropping into the water, faeces land on the shelf in a coil before being flushed away. This is reflected in the shape of the plastic variety sold in joke shops. Then she hands me a carrier bag and warily I put my hand in. There's a used toothbrush, a jar of green slime, barbecued worm crisps, a flat pool of plastic vomit, a model finger which oozes a yellow substance when you squeeze it, a severed plastic hand with a bloody stump and finally whitish-yellow pustules to stick onto your face. Val Curtis researches disgust and hygiene for a living so she does have a good excuse for owning all these revolting things.

The contents of this plastic bag represent the main categories of object we find disgusting. All bodily products with the exception of tears seem to revolt us, and the idea of pus, vomit or any of these products reaching our mouths is especially powerful. It was notable that at the most revolting moments at the public autopsy in London, members of the audience covered their mouths with their hands.

In an attempt to discover the purpose of this strange emotion Val Curtis collected data on the things that most disgust people across six different countries. While there's been much discussion in the scientific literature on the differences between cultures – one finds cheese disgusting, another insects – it was the similarities which intrigued Curtis. People repeatedly came up with the same list: bodily excretions, body parts, decayed or spoilt food and anything slimy. But it was the other part of Curtis' work – the study of infectious diseases – which provided the clue as to the purpose of disgust. While researching the prevention of disease through improvements in hygiene, Curtis happened to look at a list of disease-carriers in a textbook – vomit, perspiration, breath, faeces, mucus, semen, vaginal fluid, decaying matter, saliva – when it occurred to her that it was the same as the list of things which people found the most disgusting. Everything on the disgust list had the potential to carry disease. She concluded that we must have evolved to feel disgust in order to avoid disease. Just as we've learned to fear dangerous animals, so those with a strong sense of disgust were more likely to survive for long enough to breed. It's essential that anything disease-carrying is not allowed to enter our bodies.

There's a phrase often used within disgust research that repels me just to hear it – the body envelope. The idea is that anything that crosses the boundaries of the subject's body can give rise to disgust. Thus if dirt gets into the mouth that's a breach of the body envelope. Overpowering feelings of disgust help us to avoid this. Things we find disgusting often look or smell foul to us, but if we touch something disgusting it adds that extra dimension. I remember walking along a road in India wearing open-toed sandals when I heard a loud sneeze from a man beside me. I felt something warm and wet on my foot and when I looked down there was a big blob of yellow mucus on my big toe. Other people's bodily substances are so much more revolting than your own. I ended up rinsing my foot in a warm dirty puddle in my determination to rid myself of the mucus. Again it's a case of keeping polluted substances away from your body. Just one gram of faeces contains 100 million viruses and over a million bacteria, so it is useful to find it disgusting. Poor sanitation always leads to the spread of disease and to deaths from infectious diseases. The Romans built the great sewer in Rome, not because they understood the link between poor hygiene and disease, but because they couldn't stand the smell. It seems their sense of disgust was protecting them.

The problem with this protective disgust hypothesis is that, as with the unfortunate first date, our feelings of revulsion aren't always rational. Why be disgusted by the idea of eating some perfectly healthy cat flesh? This theory also fails to explain cultural differences. Why is a side-order of insects repulsive to some and a delicacy to others? Paul Rozin does an experiment where he gives people a carton of apple juice and a brand new bedpan. He takes the bedpan out of its sealed box, half-fills it with apple juice and asks people to drink it. Most people won't. The association with urine is just too strong. This demonstrates how easy it is to provoke irrational disgust.

Val Curtis argues that it's exactly this excessive sensibility to disgust that prolongs our survival. If just one mistake could lead us to eat something poisonous, then it's safer to cut out that whole food group, provided there is some alternative food available. There is a sense in which an item can remain disgusting even when it has been cleaned; you probably wouldn't want to use the old crockery of someone you knew to be lax about cleanliness. In our minds the essence of contamination remains. Rozin believes the reason we're disgusted by the apple juice in the bedpan is that when we were evolving we had to develop a shorthand for what we saw. Thus if something looks like a tiger you should run. Nowadays if you live in the West and see a tiger the chances are that it's in a cage, in a photograph or on television, in which case you don't need to run. Likewise with disgust. If it looks like urine in a bedpan then disgust kicks in because in most circumstances it would be a bad idea to drink a clear yellow-tinged liquid that you found in a bedpan. Participation in a psychology experiment is a rare event for which evolution can't plan.

It is interesting, however, that we evolved to experience an emotion as specific as disgust. If we need to keep away from rotten foods, why aren't we just afraid of them? Perhaps a more specific emotion than fear is needed. Fear of mould might cause you to leave your kitchen worktops covered in the stuff because you dare not go near, whereas disgust with mould propels you remove it, whilst ensuring it does not reach your mouth as you do so.

Possibly more so than other emotions, disgust is influenced by culture. At different times and in different places people are repulsed by different things. In a Roman amphitheatre watching someone sustain horrific injuries counted as entertainment, while witnessing the infliction of the same injuries today would be considered a trauma. If disgust is all about protecting ourselves from disease, it's strange that sometimes it's the deliberate defeat of disgust which promotes safety. Backpackers' guides, for example, tend to recommend accepting drinks from strangers abroad only if the stranger drinks from the cup first. In this instance it is necessary to overcome your disgust at another person's saliva in order to ensure what you're drinking is safe. Likewise dead bodies disgust humans the world over. Is this associated with fear of our own mortality or a simple way of keeping us from the infections which dead bodies might harbour? Intriguingly, unless death was caused by infectious disease, most dead bodies are harmless, even when decomposing. So it concerns Curtis that after a disaster a priority is always put on clearing away the bodies for the sake of hygiene. She believes a greater number of lives could be saved if ongoing sanitation for the survivors was prioritised. In this instance our feelings of disgust could cost us our survival, which would seem to go against Val Curtis' theory that disgust is purely an evolutionary protective mechanism, unless this is an exception. Perhaps disgust is our best overall hope for protection against disease, but with large disasters it's not so effective. It has to be remembered that an emotion can still be adaptive long-term without every experience of that emotion promoting survival. As we'll see later, even emotions that appear negative like anger or jealousy are sometimes useful.

Paul Rozin believes there is more to disgust than self-protection from disease. He says we are disgusted with ourselves as animals. Therefore anything which reminds us of our animal roots repulses us, including bodily functions. We even try to distance ourselves from the fact that we eat animals, referring to pig as pork and packaging meat in a way that helps us to forget it is an animal. Finding feathers or animal hairs on our meat can be abhorrent. As it happens, the very fact that we do experience disgust separates us from animals, none of which exhibit feelings of disgust. Just recently I watched as my friend's baby was sick on the kitchen floor and her dog ran forward and eagerly lapped it up, apparently delighted with this surprise snack.

So the fact that, like this dog, other animals don't feel disgust allows us to feel that we are different, that we have finer sensibilities which animals lack, confirming to us that we are not ourselves animals, that we are special. However there is one certainty we share with all other animals; death, the ultimate reminder that we too are animalistic. Rozin believes that a fear of death can explain some feelings of disgust. Stepping on a decaying rat in the street reminds us that we're mortal, that we too are vulnerable animals with blood, flesh and pus and that we will eventually reach the same state as the rat. This could also explain the haste with which dead bodies are removed from scenes of disaster.

matter out of place

Others approach disgust from an entirely different perspective. The renowned anthropologist Mary Douglas spent years studying the Lele tribe in Africa. Here she noticed that the people were fastidious about food rules; women, for example, would never eat chickens because they come from eggs, which would be like eating her own children. Douglas concluded that societies have a need to categorise animals and objects. In doing so people develop systems that explain what belongs where. Each member of a society is expected to make the same categorisations otherwise the system is threatened, and every society has its own rules.

When I went to meet Mary Douglas at her house she offered me a drink, telling me that the day before a friend had turned down the offer of a glass of wine because it was only eleven in the morning and drinking wine at that time would be wrong. After lunch the friend said it was too early for a cup of tea. She had taken on her society's rules governing acceptable times for consuming certain drinks. Even a nice drink can seem unpleasant when it doesn't fit in with the rules. Imagine drinking a glass of orange liqueur first thing in the morning. The very idea can make you feel sick.

Mary Douglas also noticed that the foods that the Lele people found disgusting shared a particular characteristic. She was sitting beneath some palm trees when she heard rustling overhead. A man told her it was flying squirrel, but that it was an animal they would never eat. 'Is it a forbidden food?' she asked. It was not, but it was considered disgusting because it could both move along the ground and fly through the air, making it neither an animal nor a bird. This led Mary Douglas to the idea that we have a need to categorise our surroundings and once something crosses categories or is found in the wrong place, it becomes disgusting.

flesh on the banister

My friends were renting a cottage on the steep hill that led down to the island's harbour. Not far away was the hotel where I was working as a waitress – the hotel that served up fennel from the bin. One night we decided the cottage would be a good place to have a party, so after the pubs shut we invited people back to the house. It was all going well until a man went upstairs to use the bathroom and then decided to make an entrance by sliding back down the spiral banister into the living room. We all cheered as he slid, but just as he was about to leap triumphantly off the end the last strut came loose, leaving him dangling, suspended by the back of his jeans, much to everyone's delight. He was left hanging there while we laughed at his cartoon-like suspension in mid-air, until we noticed that the one person not laughing was him. He was in agony. The strut had punctured his left buttock and was in fact impaling him. Horrified, we lifted him off and found someone with a van who could take him to the cottage hospital for stitches. Returning to the house we sat back down on the sofa, contemplating what we'd just seen. Then we saw something else. Resting on the end of the banister was a cube of flesh, rather like a piece of raw meat ready for threading onto a kebab. No one wanted to touch it, but it had to be moved. A flannel was fetched from the bathroom and one girl steeled herself to grab the flesh and the whole flannel was thrown into a black bin bag. It was the idea of touching the flesh that was so disgusting. Despite the severity of the poor man's injury, which left him limping for weeks, what everyone remembers most vividly is something he never saw; the flesh he left behind, contaminating the banister. As part of his body it was fine. Even if we'd seen it in an operating theatre it would have been all right, but the fact that it was sitting on the end of a white wooden banister in a cottage was truly revolting.

This fits in with Professor Douglas' idea that something is considered disgusting if it is matter out of place. The rotting carcass of a dead mouse in the undergrowth is not nearly as foul as the same mouse on your pillow. If a chicken drumstick is on a clean plate in a fridge then the chances are that it is safe to eat, but if you find it in a dark corner of your car-boot you would guess that it is likely to be neither clean nor fresh. In an instant the fact that the chicken is in the wrong place provokes disgust. Context is everything. Taking this a step further, Douglas believes that the reason we categorise is to achieve a shortcut for working out what is edible and what isn't. It's faster to exclude a whole category of foods like insects, than to work out which ones are nice to eat. Douglas believes that hygiene is simply a by-product of this classification. Something is abhorrent if it brings up contradictions, so eating vomit is disgusting because it involves ingesting a substance which belongs inside you. Once it's outside your body it's matter out of place. Anything that crosses categories is disgusting, so faeces are disgusting not because they spread disease, but because they come from inside the body, yet are visible outside it.

However, not all creatures we find disgusting are ambiguous in terms of their grouping. Paul Rozin gave me the example of the cockroach – an archetypal insect in that it is crunchy and six-legged. There's no mistaking that they are insects but for some reason most people find them far more unpleasant than ladybirds. Those who believe that disgust is present to prevent the transmission of disease would argue that the reason cockroaches are disgusting is that they can carry disease and often signal a lack of cleanliness.

Cockroaches are one exception, but many disgusting things do fit in with Douglas' theory. Soil outside in the garden is fine. Soil on the dining room table is dirty. You could argue that rotting food is disgusting because it's neither food nor earth. A foot of hair growing from the crown of your head is fine – beautiful, even – while a foot of hair growing from the middle of your cheek, while presenting no disease threat, would not get quite the same reaction.

Is it possible that these different theories could work together? Disgust as matter out of place could be a useful evolutionary shorthand for working out how to avoid disease. If everything stays in its place it cannot contaminate you or your food, but something that crosses boundaries, like faeces, might. A by-product is disgust at perfectly harmless things like flying squirrels or hairy cheeks, but provided you can still function then a slight over-sensitivity to disgust isn't a problem. It is better to avoid a food that transpires to be harmless than to make one mistake and catch a life-threatening disease.

the fascination of the disgusting

High up on the shelf there's a wax model of a woman with a brown, ridged horn coming out of her forehead and curving down in front of her face, the tip level with her chin. She's called Madame Dimanche and she lived in Paris at the beginning of the nineteenth century. When the horn started growing she was twenty-four. By the time she was eighty it was ten inches long.

A group of American high school girls gather around,

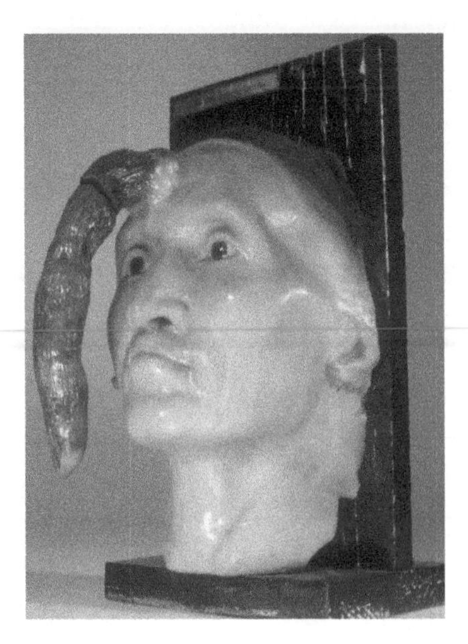

both fascinated and horrified at poor Madame Dimanche. The Mütter Museum is tucked away in a side-street in Philadelphia, far away from the city centre. It describes itself as a museum of over 20,000 objects including 'fluid-preserved anatomical and pathological specimens, medical instruments, anatomical and pathological models, items of memorabilia of famous scientists and physicians, and medical illustrations'. In truth, it's a museum of disgust. There are informative displays demonstrating the problems of cross-contamination between raw and cooked products in a kitchen, but the visitors are more interested in staring at the grey-green human hand kept in a cylindrical jar. It was amputated after frostbite had turned the limb gangrenous.

Walking through the museum, you find your eyes darting around, looking for an exhibit fouler than the last. In an extra-large jar there's a human back and shoulders with what looks like a huge red-raw hole gashed out of the middle, criss-crossed by yellow snakes of pus. It turns out to be a carbuncle 'in the healing stage'. I dread to think what it looked like before it began to heal. The most fascinating thing I saw looked like a metre-long, crispy, grey elephant trunk, only it was fatter. It had been taken from the intestines of an unfortunate man who, in the black and white photograph taken before the removal of the growth, appeared to be nine months pregnant.

For some reason fascination appears to accompany disgust, almost transforming it into a positive experience. We enjoy testing ourselves to see how much we can stomach. I once recorded a prostate operation for a radio programme on anaesthesia. Although watching made me wince, I was intrigued by the whole experience. A camera filmed the inside of the man's abdomen and, as it played out on the screen you could see white chunks floating about, like pieces of bread in a duck pond. The surgeon told me it was 'matter' and that by the time a person is in their forties, there's plenty of harmless debris floating about. They even sucked some out while they were in the vicinity.

Paul Rozin calls our fascination with the disgusting 'benign masochism'. We get the chance to do something that feels threatening, but is in reality safe. A disgusting exhibit in a museum allows you to experience a powerful feeling without exposing yourself to any danger. As an academic whose subject area is disgust, Val Curtis is often told that it must be hard for her to get dinner invitations. In fact, when she goes for dinner her friends begin the evening saying that disgust must not be mentioned and then spend the entire evening discussing it. When the novelist Chuck Palahniuk was holding readings of his famously disgusting short story, *Guts* (a first person account of a boy whose onanistic experiences in a swimming pool go horrifically wrong with permanent consequences), he found the venues were packed. The tales of people fainting in horror at the readings (forty so far according to Palahniuk) didn't seem to deter people; it made them even keener to hear the story. And we happily watch the television with delicious horror while contestants in reality shows are forced to eat live stick-insects or drink pints of sick. However, disgust is an obsession that's far from modern. If you go into a cathedral and look under the chairs in the beautifully-carved choir stalls, amongst the devotional tableaux of praying monks and the mythological scenes of animals fighting, the occasional misericord simply depicts the downright disgusting; such as the men with legs splayed wide, clearly defecating. There seems to be a connection across the centuries between the medieval craftsman and the *Viz* cartoonists of the late twentieth century – an amusement at anything disgusting related to bottoms. Shakespeare also dwelt on disgusting details. Take *Henry V.* Shakespeare provided the king with plenty of rallying speeches about the nobility of war, but could not resist some revolting extras:

> And those that leave their valiant bones in France,
> Dying like men, though buried in your dunghills,
> They shall be famed: for there the sun shall greet them,
> And draw their honours reeking up to heaven,
> Leaving their earthly parts to choke your clime,
> The smell whereof shall breed a plague in France.

We seem to have a particularly good memory for revolting events. I can never remember jokes but when it comes to disgusting stories I still can't forget the hopefully apocryphal story of the couple whose hotel room had been ransacked, but nothing seemed to be missing. Even their camera was sitting where they had left it on the dressing table. They soon forgot about the break-in until they returned home and had their photos developed. There, amongst the snaps of beaches and bars, was a photograph of two bare bottoms, protruding from which were their own toothbrushes, the same toothbrushes they had been using for the past fortnight.

When I've asked people to recount their stories of different emotions they remember the disgusting stories far more easily than they do the times they felt angry, let alone joyful. One friend told me about the time he had dysentery and unable to control himself, had messed his underpants. He went to some public toilets, cleaned himself up and wanted to throw away the pants, but there was no bin and he didn't have a bag with him, so he hid the pants behind the cistern and left. Within minutes he felt so guilty about the cleaner who would find his foul pants that he begged a carrier bag from a shop and headed back to the toilets. When he returned the place was still dirty, so no cleaner had visited, but his clothing had disappeared. He is still haunted by the feelings of disgust he experienced at the idea that someone had wanted his soiled pants.

Our good memory for the disgusting can lead to a person's whole identity becoming marked by one revolting incident. A few years ago an American businessman was in the news after he was refused more alcohol on a plane and lost his temper. He raged through the plane, threatening the staff, but what gripped the public's imagination and earned him the nickname 'Jet-mess Exec' was that he climbed on top of the drinks trolley and defecated on the top shelf. He then cleaned himself with linen napkins and wiped his hands both on the trolley and on nearby seats. In time people will forget that he was abusive towards the cabin crew, but everyone will remember what else he did, even though that behaviour was less frightening for the individuals concerned. Even the judge, sentencing him to two years' probation and a $49,000 clean-up fee, noted the degree of humiliation he had experienced.

Just one revolting scene in a film can identify that film as forever disgusting. *There's Something About Mary* is famous for one short scene where a woman turns up for a date, sees a substance dripping from the man's ear, assumes it must be hair-gel, scoops some up and puts in into her hair. In fact it's semen which slowly dries in her hair, creating a crispy quiff. The moment where she puts the 'gel' in her hair lasts only seconds, but this has become the best known element of the entire film.

disgusting fun

A man walks into a joke shop (and no this isn't the start of a joke) and asks if they sell fake dog faeces. The shop assistant takes two off a shelf, one coiled and one straight and puts them on the glass counter. 'This one's great,' says the man, selecting the former, 'I need another one exactly like it.' The assistant puts a third on the counter. The man peers at it closely, turning it round in his hands to examine it in more detail. 'It's not quite the same. This one's more textured at the tip.' Half an hour later there are more than thirty lumps of brown plastic laid out in rows on the counter, each varying slightly in shape or shade. A mother who's come to buy her son a new whoopee cushion after his was confiscated at school looks askance at this man taking such care in his examination of the dog dirt. Little does she know that the customer is a world authority on disgust, who must have identical joke faeces to maintain reliability in his experiments. He's Paul Rozin, whose theories on disgust I've already mentioned. What struck me when I met him was his enthusiasm for the subject. He loves it. Part of the reason is the entertainment value of asking people to do unpleasant things. Even sourcing the disgusting items is an adventure: 'We had to find a butcher to give us a severed pig's head. Then we gave it to people and asked them, would they look at it, touch it, put a pin in its eye? We had a Nazi officer's hat. Would they look at it? Would they touch it? Would they agree to put it on? Since we think death is important for disgust we wanted to use cremated ashes. Everything else we used was real, but we were a little queasy about getting human ashes so we decided we would get something that really looked like them. We weren't quite sure what they looked like so I went with my student to a crematorium and we found a guy and asked him to show us. This old fella was so pleased to see us. He must have been in his seventies and he said we were the first people who had expressed an interest, so he showed us box after box of ashes, telling us how sometimes they have a little blue in them. He opened up a whole world for us and even asked whether we wanted to watch bodies burning. In the end we used bonemeal and added a few things to it. Researching disgust takes you into all these adventures. We bought Japanese candy locusts to see whether people would eat them. We even had people watch a movie of people eating a live monkey brain. They could stop the film any point and we timed how long they watched it for. It was great fun.'

Rozin clearly sees the funny side of researching a subject like disgust. In addition to its fascinating side there is something amusing about the revolting. I know one man who as a student would mix together butter, cocoa powder and icing sugar, smear the mixture on the bottom of his trousers and wait for a friend to remark that he must have stepped in something unpleasant. 'Are you sure?' he'd say, scooping some up with a finger and sucking his finger clean, while everyone else watched in horror.

Could our amusement at disgust be a way of dealing with its horrors or perhaps another way of underlining what is and is not acceptable? This might be why children are so interested in the foul; they are learning where the boundaries lie. I used to be particularly fond of the revolting rhymes my father told me. This type of poem was created by Harry Graham back in 1899, but has been imitated ever since. Examining the contents now, the poem is in fact more unpleasant than many a horror film, but at the same time it is funny:

Willie with a lust for gore
Nailed the baby to the door
Mother she began to faint
Saying 'Willie dear, do mind the paint!'

Little Willie, wicked sprout
Gouged the baby's eyeballs out
Stamped on them to make them pop
Mother sighed, 'Now, William, stop!'

Children in particular are often fascinated with disgust. In the Science Museum in London there's one case in the Space Technology Exhibition around which children always cluster – the case containing an astronaut's underwear along with an explanation of how people go to the loo in space – something we all want to know.

When I interviewed a group of eight-year-old schoolchildren about disgust, they were wary at first of telling me what they find repellent, afraid that they might get themselves into trouble, but when I reassured them they could say whatever they wanted to, with delight and giggles they told me about the most revolting things they could think of. Parents sometimes worry about their child's preoccupation with the disgusting, but in fact this fascination seems to be part of the normal developmental process. Children come to terms with the rules of disgust by taking what might appear to be an unhealthy interest. They are also quick to learn that mentioning or doing anything disgusting is a superb way of guaranteeing a reaction from the adults around them.

The allure of disgust contributes to our survival by focusing our minds on the disgusting and reminding us of what is safe and what is dangerous. We might watch with horror as somebody lies in a coffin of rats on a TV show, but this underlines the fact that normally rats are to be avoided.

condemnation through disgust

Revulsion at bodily products or fetid food is not the only type of disgust that children learn. Through disgust parents can convey attitudes towards sex and death and even towards people from other groups. Disgust is such a powerful vehicle for teaching that it's not hard to harness it for harm.

Everyone enjoys with vicarious horror a good story about someone going abroad, sitting down as a dinner guest and being served with eyeballs or a live fish flapping on the plate. While it is undoubtedly entertaining, disgust can be used as a subtle way of criticising another culture. 'Look what they do – if they're prepared to do that they must be revolting.' Disgust can soon spread from something that threatens our body to something that threatens our very being, as Paul Rozin puts it, 'A mechanism for avoiding harm to the body became a mechanism for avoiding harm to the soul.' Disgust at the behaviour of other groups can also be manipulated. In July 2003 under the headline 'Swan Bake' the *Sun* printed a story claiming that asylum seekers were stealing the queen's swans for barbecues. Eventually the *Sun* printed an apology admitting that there was in fact no evidence to suggest that the swans had been taken by asylum seekers, but it shows how accusations of disgusting behaviour can be used to condemn an unpopular group.

Primo Levi, the chemist who became famous after the publication of his extraordinary chronicles of the Holocaust, described his journey by train to the concentration camp. No latrine was provided in the train. Although the Nazi soldiers told the prisoners to bring money and valuables, no one advised them to bring anything to use as a toilet for the long journey. At an Austrian railway station the prisoners were allowed to disembark the train, but could not wander far, leaving them no alternative but to defecate publicly, conveniently providing the Nazis with the perfect opportunity to demonstrate how disgustingly Jews behave. Once this dehumanisation through disgust had begun, their appalling treatment could be seen as more acceptable. Throughout history, features of disgust like dirt or offensive smells have been used to castigate other groups. Not only do we lose compassion for a person once we feel disgusted by them, but we seek to distance ourselves. According to the law professor, Martha Nussbaum, seeing a Nazi officer as someone who is disgusting helps you to believe that you could never be someone like that yourself. If you portray paedophiles as disgusting then you can start to see them as closer to animals than other people and it removes the need for compassion. Both misogyny and homophobia employ the same tactic. Just one behaviour can dehumanise members of a group. Take cannibalism – what could be more disgusting and condemn people more instantaneously than eating other people? In 1979 the American anthropologist William Arens challenged popular ideas of cannibalism, declaring that tales of human-eating among certain societies were a myth, created with the aim of denigrating certain tribes. Soon afterwards papers were published demonstrating that cannibalism had existed previously, but was often only used in a symbolic sense as part of rituals. One tribe that did practise cannibalism in the twentieth century was the tribe of the Wari people from Brazil, but far from capturing enemies and boiling them in giant cauldrons, the consumption of dead bodies of people they knew well, particularly children, was seen as an act of compassion. The bodies were roasted with the greatest respect and the closest relatives didn't take part. In the 1950s and 60s government-sponsored expeditions sought to stop the practice, forcing them to bury their dead instead. Then it was the turn of the Wari people to feel disgusted because the ground was considered cold, wet and polluting, a revolting place in which to leave a person you loved. Beth Conklin has studied the society and says that even recently people looked back with nostalgia on the time they used to eat their dead, saying it helped them to come to terms with their grief. This bore no resemblance to the way cannibalism was portrayed in the West, where disgust has been used to condemn whole cultures.

Although our feelings of disgust can be exploited to turn us against people, it is an emotion that is on the whole beneficial. Whether it is keeping us safe from poison, providing us with entertaining tales to tell or even helping us to avoid anxieties about our own death, disgust is an emotion we appear to be hard-wired to develop. As we grow up we learn the items of disgust peculiar to our own culture, whilst also realising that disgust at matter which is out of place can provide a shorthand for the fact that something is disgusting and should be avoided. The slight oversensitivity that we have when it comes to disgust causes us to find many harmless creatures and foods repulsive, but it is essential; just one mistake could be the difference between life and death. Most of the time we can override the feeling when absolutely necessary. Disgust only becomes a problem if it is so strong that a person is left unable to deal with everyday life, for fear of contamination. The rest of the time we should rejoice in disgust.

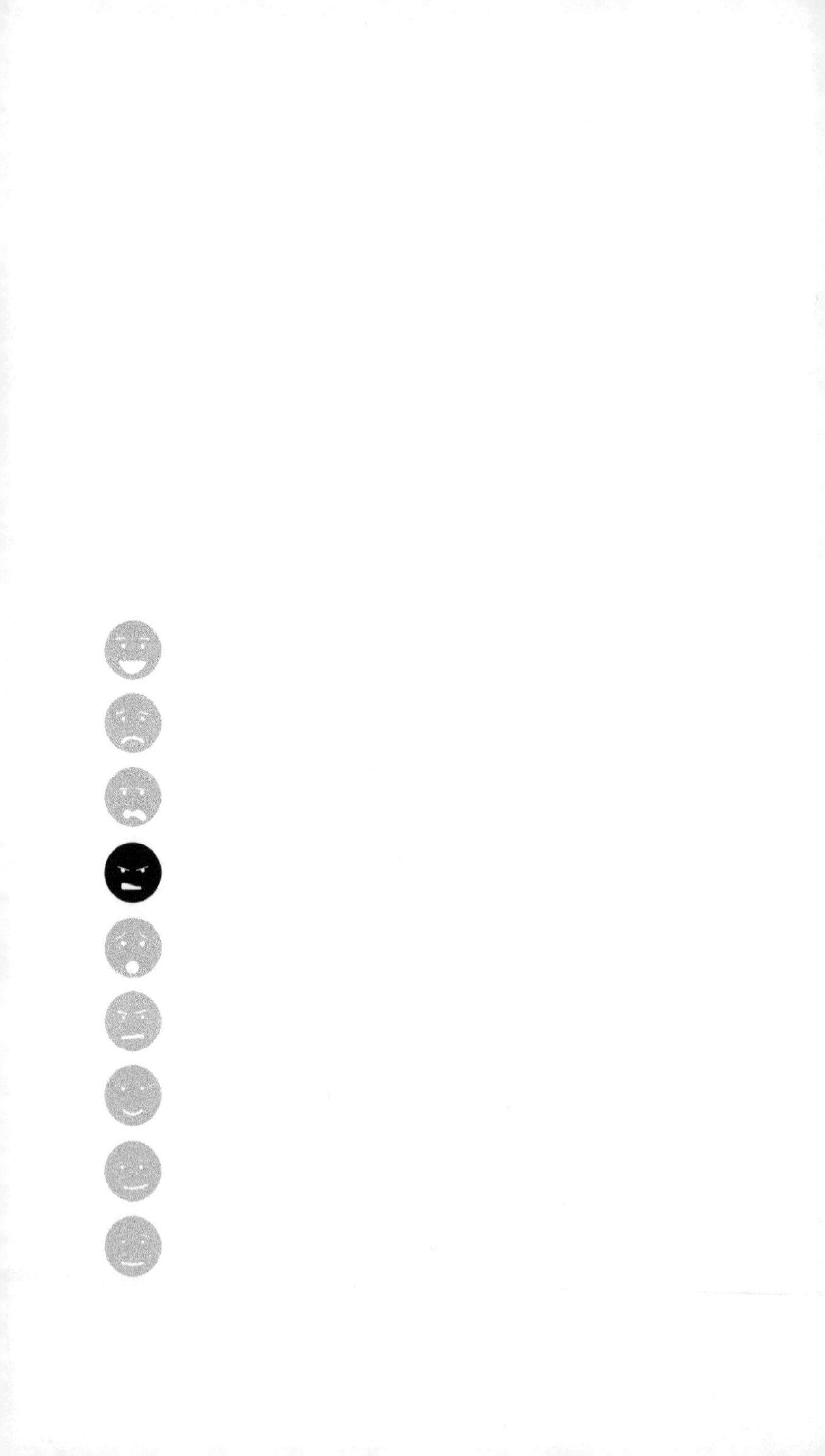

four

anger

Leigh Ann Barton was proud that her husband Mark had a steady career as a chemical salesman in Atlanta, USA; a job which earned him enough money to dabble in stocks and shares. She knew that recently he'd had a run of bad luck with his investments, but she had no idea of the sums he was gambling – and losing. In just seven weeks in the early summer of 1999, Mark Barton lost more than $100,000 on the stock market. When she found out she was furious. She was shouting and screaming; he had ruined their lives, he had destroyed their future, he was selfish, stupid, thoughtless, an idiot. How could he be so stupid? How could he be so thoughtless? What about the children? What about her? He had never seen her like this. And then it was he who snapped. How could she speak to him like this? It wasn't his fault that the shares had gone down. Who was running these companies? What were they up to? Here he was, trying to do the best for his family, to make them some money so they could have the new computer games they wanted and go on nice holidays like their friends. All this effort and he was the one getting shouted at. He looked around him and saw his claw-hammer lying on top of his tool box. Something filled him from within. He felt a surge of power. He picked up the hammer and smashed it into her face again and again.

In that moment of intense anger the family man had become a killer. He hid her body and told nobody. The most normal actions were now calculated actions. The next day it was back to family domesticity. He dropped his eleven-year-old son off at scouts and then spent the afternoon playing with his seven year old daughter. That night he waited until the children were asleep – the natural actions of a loving father, except that he was biding his time. When he was sure they were sleeping soundly he took out the same hammer he had used on his wife, tiptoed into their rooms one at a time and bludgeoned both his son and daughter to death. He didn't want them to feel any pain, so he filled the bath with water and put each of them face down under the surface to ensure that they were dead. Then he dried them off with a towel, an action so familiar, but that day so different, and put them back into their beds for the last time with their favourite things beside them – a cuddly toy for her, a video game for him. But this is not the story of a father who kills those he loves in order to spare them the pain of losing their mother. For the next day the domestic killer became a random gunman, visiting the offices of two firms of stockbrokers and opening fire, killing the nine people who happened to be there at the time – a grocer selling some stocks, the office manager, a mother taking a computer course. Even then Barton didn't surrender and when the police caught up with him a few hours later he shot himself, leaving behind three notes and a letter. One note was full not of remorse but of self-justification and self-pity. He wrote that he had vowed to kill 'the people that greedily sought my destruction'. This killing spree was not the result of one instant where he lost his temper. His rage had built up to the point where he committed murders over three separate days. The killings did seem to follow on one from another, however. It was a complex web of rage. His children were from his first marriage to a woman who had been hacked to death not long after he had taken out an insurance policy on her life. The insurers had refused to pay out the full amount, suspicious that he might have had some involvement in her death. Unable to prove it, they had come to a settlement. This was the money he used to begin his dealings on the stock market. After murdering his second wife he decided to kill his children to save them from growing up without a mother or father. Once he had destroyed his family, he then turned on the financial centre where he'd lost the money that had begun the row with his second wife.

This form of rage – not only violent, but enduring – is pretty rare. Most people will never experience it or, if they do, they are unlikely to act on it. Karl experienced a far more everyday sort of anger than Mark Barton when he was in his twenties: 'I was angry all the time. I was the epitome of the angry young man which occasionally led to violence. It was almost like a background dissatisfaction which has a tendency to come to the fore and can be misplaced. So it's not that you'll have a rage over a particular event, more that there's a kind of background stuff that you're not dealing with that will come to a point of explosion.'

Years later he thought his angry days were behind him, when he and his partner split up and suddenly the feelings of rage returned. 'I found myself going back into a sense of rage I hadn't experienced for years. It came as quite a shock to me that all the controls I thought were in place were only actually so deep. I would actually drive off in my car or leave the house, anything to avoid a confrontation. I have a belief that a confrontation will always end in violence, so if you want to avoid the violence you avoid the confrontation.'

In fact tempers lead to aggression less often than people tend to think. Sometimes anger can even halt a confrontation. If a parent stares crossly at two fighting children they might stop, or an angry reaction from a victim of bullying might let the aggressors know that it's time to cease their taunts. Although people might appear out of control when they are angry, in fact they do tend both to take the situation into account and to abide by their society's rules. So although a person might feel murderous with fury, only rarely do they take it that step further.

Anger is a particularly tricky emotion to study because it is so hard to predict when somebody is about to lose their temper. Even if you were to wait for a hot day and then sit observing drivers sitting in a traffic jam, it could be hours before you saw anyone explode with fury. Moreover, most episodes of anger take place in private where they are even harder to study. Instead researchers have tended to opt for studying anger in the laboratory using unsuspecting participants. The usual procedure is to give them a task and then arrange for someone else to thwart their progress in some way, perhaps by talking incessantly or by interfering with the equipment. Later they have the opportunity for revenge when they are given the job of teaching the irritating person lists of words while doling out small electric shocks as punishments for mistakes. The idea is that the angrier the person feels, the stronger the shocks they will choose to give the other person. The artificiality of the situation is a problem. The participants are all too aware that their behaviour is being observed so naturally they will want to show themselves at their best. It's hard to know whether this is how they would behave in real life.

An alternative approach is to interview people or give them questionnaires about the typical situations that make them angry and how often they lose their tempers. Here, there's a different problem; although people might become angry quite frequently, many instances are soon forgotten. To overcome this, just as they did with joy, researchers asked people to keep diaries where they noted every episode of anger.

One fan of these sorts of studies is the psychologist James Averill from the United States. In his in-depth research on anger he found that 85% of people reported getting angry once or twice during the previous week. Of the 160 people in his study, one poor person became angry more than ten times each day, whilst twenty-six people didn't get angry at all during the week of the study. People were less likely to lose their temper with people they disliked than with the people to whom they felt closest. These would probably be the people with whom they were spending the most time, but crucially they were also the people from whom they expect the most, which means there might have been more to gain by becoming angry.

what's anger for?

Anger doesn't tend to get a good press, but although we associate it with tragic stories like those of Mark Barton and his family, it can be useful. Anger readies our bodies for attack, giving us that extra bit of vigour which might prove essential. No other emotion is able to keep the body at a high pitch for such long periods. This might have been particularly useful in the past.

If you lived in a small tribe and had reason to fear the neighbouring tribe, the ability to anger quickly would be essential for survival. Today, anger appears less useful. Instead of a tool of self-defence more often it's an emotion which upsets people and of course occasionally leads to violence.

The philosopher Seneca wanted anger banished long ago. Between AD 40 and 50 he wrote what is thought to be the first complete work on anger where he recounts endless tales of cruelty perpetrated in the name of fury. For example the King of the Persians was so incensed that he cut off the noses of the entire population of Syria, apparently leaving the country with the nickname 'land-of-the-stump-nosed'. Seneca disapproved of the idea that anger could be a valuable emotion in battle. Instead he believed it caused men to behave rashly. In terms of the energy or bravery it might provide, he argued that drunkenness would be just as effective. 'It makes men forward and bold, and many have been better at the sword because they were the worse for drink. By the same reasoning you must also say that lunacy and madness are essential to strength, since frenzy often makes men more powerful.' Analysing angry outbursts in more detail, he concludes that the problem is that they begin with a surge, which means that, even when violence is justified, those who are nearest will suffer the most. Later on tiredness will make the person less harsh, leading to an unfair distribution of punishment.

self-defeating rage

Greg Rusedski was two sets down in a match against the number five seed, Andy Roddick, at the Wimbledon tennis championships in 2003. It looked as though Rusedski was unlikely to make it through to the third round until he had a winning streak in the next set. As he waited to receive Roddick's serve he was leading 5–2, about to win the set. Then came a momentary incident which was to so enrage Rusedski that he would throw away the match. In the middle of a point a member of the audience did a very good impression of a line judge shouting 'out'. Assuming the call was genuine and that he had already won the point, Rusedski hit the ball half-heartedly and turned away. His opponent played on, winning the point. Rusedski demanded they replay the point and when the umpire refused he began yelling in one of the worst outbursts of swearing ever heard at Wimbledon. 'I can't do anything if the crowd fucking calls it,' he yelled at the umpire. 'Absolutely fucking ridiculous. At least replay the point. Fucking ridiculous, fucking ridiculous, frigging ridiculous. Some wanker in the crowd changes the whole match and you allow it to happen. Well done, well done, well done. Absolutely shit.'

Anger diverts energy towards the muscles and, in contrast to other emotions, this is not yet momentary; the increase in energy is maintained. Many sportspeople use this to their advantage, deliberately stoking up their anger towards the other team. John McEnroe was famous for shouting at umpires and flinging his racquet down onto the court. Having said that, surprisingly for a man who was supposedly the most angry man in tennis, the worst McEnroe ever called an umpire was 'the pits of the world'. After his outbursts McEnroe would convert that extra energy into fantastic winning shots against a now distracted opponent. In contrast, Rusedski's anger did not work in his favour; he remained so furious that he lost all concentration, and eventually the match itself. The point he lost was inconsequential, but the effect on his mood afterwards was crucial. He was out of the tournament and was fined £1,500 for swearing, a fine that the man who shouted 'out' from the crowd graciously offered to pay.

Anger might be a hangover from our past that we'd rather wish away, but it's not always been viewed as negative. Writing in the fourth century Lacantius considered anger to be so useful that it was a gift from God. Homer described it as 'sweeter than honey'. It's true that it does have its uses. Firstly it acts as a particularly strong signal to others, hence studies have shown that facial expressions of anger are especially easy to identify. If people are given a photograph of a crowd to scan, their eyes move towards angry faces faster than other expressions. However, the brain also finds anger hard to process and the scanning eyes take longer to move away from an angry face and on to the next face. For this reason it's particularly time-consuming to scan a photograph of a crowd of angry people to find the one happy face.

Anger often represents a threat so it's essential that we identify it quickly. If a parent is angry a child knows that what they have done is serious. Surprisingly even those on the receiving end of anger can also find it useful. In his extensive research on anger the psychologist James Averill asked people to think back to specific situations where either they'd lost their temper or been the target of someone else's anger. The majority of the targets thought that the anger had been beneficial, saying that they had realised their own faults or that their relationship with the angry person had been strengthened. As you might expect this didn't happen if the furious person had been particularly aggressive. Interestingly, the people on the receiving end felt that the anger was excessive three times as often as the angry people themselves judged it to be. However the force of this finding might be reduced by the fact that this study wasn't conducted with pairs of people describing the same incident from different perspectives; each person is describing a different event.

Anger can provide the spur for people to campaign against the injustices they see. Not long after his outburst at Wimbledon, Rusedski made use of a slow-burning anger which gave him the energy to refute claims that he had taken illegal drugs after failing a drugs test. Now he's trying to change the system so that other tennis players don't have to go through the same ordeal of unjust accusations. This time, anger is working for him.

It seems that the key is to select the right moment to express your anger, or so a cautionary tale recounted by Seneca would suggest. It's the story of Harpagus who made the grave mistake of giving the King of the Persians some unwelcome advice. To punish him, the King had Harpagus' own children killed and cooked and then served to Harpagus at a banquet. While poor Harpagus ate, the King enquired as to how he found the cooking. Somehow Harpagus resisted the temptation to fly into a rage and instead replied, 'At the King's board, any kind of food is delightful.' Pleased with this response the King didn't force Harpagus to eat the rest of the flesh, but after the meal he did order the children's heads to be brought in and asked Harpagus how he found the entertainment. By controlling himself once again Harpagus was free to sneak back that night when the King was alone and wreak his revenge with a sword.

Perhaps a third philosopher, Aristotle, had the right idea when he said that the key to anger was to feel it at the right time, with the right people, about the right things, with the right motive, in the right way. Easier said than done, of course. Aristotle stressed that becoming angry is just a feeling, not something worthy of praise or blame.

when anger begins

As with all the emotions, no one agrees over the exact age when babies first experience rage. A baby's cry could be interpreted as distress, sadness, fear or anger. You can see what you want to see. To overcome this problem researchers have developed strict coding systems for measuring babies' behaviour and facial expressions. Using these techniques they have found that anger seems to emerge at between four and six months. If you hold down a baby's arms at this age, not surprisingly they become angry.

When babies are very small their only response to something painful like an injection is to put all their energy into screaming as loudly as possible in the hope that someone else will help them, but as they get older they begin to express anger rather than simple distress. This is adaptive because when they're tiny their only hope is that someone else might rescue them. As soon as they're slightly older they begin to learn what it is that they want to happen and that they can influence the people caring for them. Then the ability to signal their anger starts to have an effect. However, some believe anger develops earlier than this, without being directed at any particular person. Professor Michael Lewis, who has written extensively in the field of developmental psychology, reports observing anger in two-month-old babies, but says it's another five months before they can target their anger.

There is an idea from Keith Oatley, whose work I mentioned in relation to joy and sadness, that emotions, although far from perfect, provide a 'ready repertoire of action'. If you feel angry you can respond quickly in a certain way, rather than having to work out a novel response to every new situation. With babies it's frustrating situations that tend to infuriate them the most. In one experiment Michael Lewis gave infants a string to pull which operated a musical toy. The babies who showed the most joy at the toy's reaction also showed the most anger when the string was detached from the toy. When the link between string and toy was reinstated they would pull desperately on the string, suggesting that their anger was spurring them to do that bit more to keep the toy going. The experimenters tried the same task with babies whose mothers had been addicted to cocaine while they were pregnant. These babies were neither as thrilled by the toy nor as angry when it stopped working. However, the key result was that not only did they show less emotional expression, but once the toy was working again they didn't increase the frequency of pulling. This suggests that they were less good at controlling their environment. Perhaps it was the strong emotions that were helping the other babies to get what they wanted, precisely the role for emotions proposed by Oatley.

Once they learn to crawl babies get angry more often, partly because their attempts are thwarted more frequently, as their carers prevent them from reaching things, and partly because they are more likely to get themselves into trouble, resulting in them and their parents becoming angry. As children grow older parents employ various strategies to deal with their rages. The tactic of ignoring anger is used far more with girls than with boys, where parents are more likely to shout back and reward a boy's bad behaviour with attention. Toddlers have been found to ire more often if parents respond to their tantrums with anger, but of course remaining calm is easier said than done. As children begin to play with other children they learn to control their anger as they begin to develop successful tactics for getting on with other people.

Clive was just eleven when he became so enraged that he took his tiny boat, designed only for harbour-use, out into sixteen-foot waves in the open sea off Carnoustie in Scotland. Earlier in the day he and his father had fallen out because his father wanted to go fishing, while Clive wanted a lift to his friend's house. When his father set off for the harbour in his car, Clive was so incensed that he followed on his bike, but however fast he pumped his knees, he couldn't catch him. By the time he arrived at the harbour, sheltered in the rocks, he could see his father's boat heading out to sea. So he decided to follow. He jumped into a pram – a tiny plywood boat for one – and because he was facing the land as he rowed himself out to sea he was oblivious to the rough conditions ahead of him. Soon he barely needed to row because the current was dragging him further out, with waves by now so vast that the horizon was constantly disappearing and reappearing. When he saw his father's boat he shouted angrily to him to come back, but slowly his fury began to be replaced by panic. With it came a new energy and somehow he managed to turn the boat and fight his way against the current and back to the harbour. By the time he was back on land he was in shock, aware that his outburst of anger could easily have ended in his death.

For Clive this sort of anger was unusual but, even by the age of five, some children are already finding it hard to control their tempers. Research has found that these are often the children whose mothers lose their tempers the most with them.

anger at work

Once children have grown up we tend to think of anger as something expressed most often between couples at home or perhaps out in the street after closing time. However, there is an arena where anger can have important, often neglected consequences and that is the workplace. As Greg Rusedski pointed out in his post-match interviews, when most people lose their tempers at work they are not watched by thousands of people all around the world. For the rest of us, the audience might be far smaller, but it can still matter. Until recently the subject of emotions in the workplace was neglected amongst most organisational psychologists. Although there has been plenty of research on stress at work, the specific role of anger has tended to be overlooked. Rob Briner from Birkbeck College in London was one of the first organisational psychologists to see emotion at work not as something which should be banished, but as something which has its place. In his research he asked people to keep diaries of any incidents where they felt angry at work. When he followed them up months later he found that if the problem had been sorted out staff could barely remember feeling cross, in contrast to unresolved anger which festered and grew. Briner gives the example of a person whose colleague has just been promoted in preference to them. They might decide it was simply bad luck or they might start wondering whether they are liked, whether this is a fair organisation and whether it might be a rather unpleasant place to work. The next time the person feels unwell, rather than struggle into work for the sake of the company and their colleagues, they might stay at home instead and even start looking for another job. Through the build-up of anger, one badly-handled situation can result in a company losing a valuable member of staff. Briner found that this was most marked when people had a clear sense of what they expect in return for their hard work. If people feel angry that the organisation is not keeping to their side of the deal they might continue to fulfil the requirements of their job, but without doing anything extra. He found that managers often avoided drawing attention to a difficult situation in case it ended in confrontation. Instead they assume the person will soon forget about it. In fact they don't. He found that if people felt positive about their work they would carry out what are known as 'organisational citizenship behaviours'. These tasks aren't officially part of a person's job; for example, clearing up some rubbish from a corridor or taking the time to help a new colleague. As soon as staff are angry they withdraw this effort. He found that one of the things that angers people the most at work is unfairness. As with siblings, if you see that others are treated better than you, you feel angry.

Few people confront their bosses about their feelings of anger, partly because they fear that they might rage too much and be seen as lacking in self-control. Psychologist Sandi Mann found that employees reported hiding their true feelings and faking emotions in about 60% of work-place communications. She insists that anger is neither good nor bad; it's how it's resolved that matters. The occasional outburst can be effective. If people think you're happy to be given all the extra tasks which no one else wants to do, they will continue giving them to you but, provided the culture of your workplace allows it, if you get angry just once they won't forget.

There are other positive sides to anger in the workplace. For a start it can motivate you to find somewhere better to work. Rather than just feeling angry with your boss, the key is to work out what the anger is telling you and to work out exactly which aspect of your job is making you feel resentful.

Anger can even be required in some jobs. This is known as emotional labour. So just as nurses need to show that they care, at certain jobs the successful people are those who can communicate their anger. A person staffing the phone lines for a debt collection agency needs to convey some irritation and even anger to demonstrate that the situation is serious and urgent. Dealing with other people's anger can be more difficult, but it's something which people working in call centres are expected to handle. By the time customers have spent ten minutes on hold, they feel frustrated even before they have begun to speak. A common tactic for call centre employees is to say they will pass the call on to their supervisor. This not only gives the customer the impression that they are receiving special treatment, but crucially requires them to contain their anger in order to explain their case again. Rob Briner says the golden rule when dealing with an angry customer is never to lose your temper yourself. Unfortunately as soon as an angry person has slammed the phone down on you, another call automatically comes through – no ringing, no control over when you answer it – the next call is there before you've even had time to turn to your colleague and tell them what just happened. Well-run centres have a set time at the end of the shift or even half-way through for people to discuss any tricky calls they've had.

There's also a second difficulty relating to anger at work and that is the problem of emotional dissonance. This is when you feel one thing, but display something else. In everyday life we constantly cover up disappointments or pretend to be enjoying ourselves while wishing we were somewhere else, but anger is one of the hardest emotions to disguise. Over time the constant suppression of anger can become stressful, leaving a person no longer knowing what they really feel.

anger and the brain

The sorts of things which typically make people angry are pain, physical discomfort, and the thwarting of behaviour through verbal or physical restraint. In these situations most people can still make a judgement as to an appropriate reaction. A few people, however, seem unable to do this. Their rage is so all-encompassing that they do things they regret. At the time I was writing this chapter, I was walking through London late one summer evening. It was still light and suddenly I heard shouting from a parked car and saw a man yelling at a woman. As I passed by he got out of the car and slammed the door so hard that the window shattered. He stormed up the road and meanwhile the woman left the car and walked away in the other direction. After a while he stopped marching into the distance and glanced over his shoulder. When he saw that the woman was neither watching him, nor taking care of his car with its broken window and running engine, he was transformed in an instant. His face went red and he became enveloped by rage. He sprinted back along the street and grabbed her roughly. By now passers-by had started to look concerned and a man leant out of window and shouted to him to leave the woman alone. He responded by threatening to send his friends around to beat the man up. He did, however, let go of the woman and they got back into the car and he sped away, wheels spinning on the tarmac, still shouting at the top of his voice. I'll never know what made him so angry in the first place, nor what he did next, but the rage did appear to possess him. The question is where this anger had come from. Had it emerged from deep within his brain?

In the Batman cartoon strip there was a character called Amygdala who was unable to control his constant rage. He was named after the amygdala, a walnut-shaped area deep inside our forebrains. In fact, it's not the case that the amygdala alone is responsible for the rage the man felt at his girlfriend. Electrical stimulation of the amygdala can make patients feel an inexplicable rage, but this doesn't happen in people who rarely become angry in real life, suggesting that as with other emotions, various brain systems come into play to form the feeling. Parts of the brain connected with thought are likely to be involved, helping us to assess a situation by integrating it with our previous knowledge of the potential consequences of different behaviours.

A few people find it hard to make a considered decision while they feel angry and these are the people whose anger is likely to land them in trouble. It seems to be very different from the well-appointed anger which might change a situation for the better. Unless they have a personality disorder, people who have extreme, explosive outbursts tend to regret what they've done and often remark that they don't know what came over them. There is a theory that the reason for the lack of control over impulses is caused by low brain levels of serotonin, the neurotransmitter I've discussed in relation to sadness. Whereas in depression serotonin levels are temporarily low, with impulsive aggression the low levels are thought to be permanent. The idea is that serotonin might delay your impulses, so even though you feel like hitting someone, you hesitate for a second, which might be just enough time for you to consider the consequences and to decide that you should stop yourself. Without that serotonin, there is nothing to hold you back. A lack of serotonin might also be involved with other impulsive behaviours like self-mutilation and bulimia.

Another theory is that serotonin somehow keeps the mood stable even when something upsetting happens. Therefore a person with low serotonin levels finds it harder to return to a neutral mood and instead responds with anger. Research on serotonin has been limited by the difficulty of measuring levels directly. It is possible to work out how much serotonin a person has by examining their spinal fluid, but this requires a lumbar puncture, hardly a procedure that's likely to attract volunteers. However, researchers have devised a clever way around the problem. There's a drug which has an effect on serotonin; as a result hormone levels are influenced and these can be measured. If people have low serotonin levels this hormonal response is delayed by about three hours. Using this method, it appears that the people with low serotonin do seem to be the people who are most prone to aggression.

In a small room at the Institute of Psychiatry in London Alyson Bond tries to make people angry by giving them competitive or frustrating tasks. She advertises for volunteers and then grades their levels of everyday aggression using questionnaires. You might think that people are unlikely to admit to much aggression, but in fact Bond finds that aggressive people are often quite happy to admit to it because they genuinely believe that in certain situations aggression is the best way to respond. In the experiment each person is given control of a piece of equipment that plays tones of varying volumes into another person's headphones. If the person in charge of the machine chooses to they can make the tone uncomfortably loud. Intriguingly, when participants' serotonin levels are artificially lowered using drugs, they show an increase both in blood pressure and heart rate and at the same time they begin delivering louder tones; i.e. they become more aggressive. Occasionally they get so angry that they tear off the electrodes measuring their physiological state and storm out of the room. Bond found, however, that this reaction did depend on whether the person was already prone to aggression. People who would never dream of being violent don't start lashing out the moment their serotonin levels drop.

If low levels of serotonin can make a person impulsively aggressive, then it follows that artificially boosting those levels with drugs like Prozac should make people behave less aggressively. This has worked, but wouldn't provide a complete solution to the problem of crime because serotonin is only implicated in a very specific type of unpremeditated aggression. It would not work for everyone who is violent. In contrast to depression, aggression can respond to Prozac after just a few days, which opens up the future possibility for those prone to impulsive aggression to take the drug from time to time when they feel that events are building up and getting on top of them.

Professor Phil Cowen, whose work I discussed in relation to drugs for depression, told me that he often ends up taking SSRIs such as Prozac as part of his work. The drug made him so much nicer that after he had been taking it for a few weeks his family asked if he could be kept on it!

The relationship between serotonin and aggression in women is weaker. One might suppose that this is because anger in general is less of a problem for women. But whilst men are twenty-seven times more likely to commit murder than women, this is anger at the extremes; there is evidence that women lose their tempers just as often as men. The difference is that if an altercation becomes physical men are likely to do more damage and are more likely to use a weapon. An alternative explanation for the differences between men's and women's aggression in experiments on serotonin is that neurotransmitters might work differently in men and women. This idea is supported by research from Stephen Gammie at Johns Hopkins University. In studies on mice he found that if the gene which produces another neurotransmitter, nitric oxide, was removed, male mice became aggressive and starting fighting with other mice. Female mice only tend to behave aggressively when they are guarding newborn pups, so the same experiment was tried on new mothers and it was assumed that if they were stopped from producing nitric oxide they too would become aggressive. It didn't happen. Instead the altered mice ignored intruders while normal mothers attacked them viciously. This suggests that this neurotransmitter is working differently in the brains of male and female mice. It is possible that the same process could take place in humans, but we're a long way from knowing; with an emotion like anger which is governed by specific social rules, it's particularly hard to extrapolate from mice to humans.

Work on violence and the brain has also looked at aspects other than chemicals. Using positive emission topography, known as a PET scan, you can see which parts of the brain are the most active by measuring the speed with which the cells use up glucose. The faster they use it up, the more active that area of the brain is. In one study they scanned the brains of murderers to compare activity in their brains with other people's. They found a lack of activity in the prefrontal cortex, the part of the brain lying behind the eyes and forehead. The job of the prefrontal cortex is to regulate and control behaviour – to make you think before you act. This is far from proven at the moment but the theory is that a lack of activity in this area might leave people with nothing to stop them from acting on their aggressive impulses.

Increased levels of the brain chemical, dopamine, which I mentioned in relation to joy, are also associated with anger in humans and with offensive aggression in a variety of species of animals. Extraordinarily, if people are given a drug which blocks the production of dopamine they become unable to recognise facial expressions of anger in others, but can still identify other emotions. It's intriguing that feeling angry and recognising anger in others both make use of the same neurotransmitter or chemical messenger. Once again this suggests that part of the purpose of emotions is to communicate information to others.

is it bad to bottle it up?

The washing was going round and round in the machine, but there was an odd sound. Something wasn't quite right. Within half an hour, loud bangs were coming from the machine and there was smoke inside. Straining to reach under the working top, I managed to turn it off at the wall. The machine went quiet and soon the smoke had gone. It was a new machine and so it was still under guarantee, but the moment when I saw the smoke turned out to be the calm part of the drama. I phoned customer services and was kept on hold, listening to tinny music while my phone bill rose. Putting the phone onto speaker, I dared to wander across the room to water my plants while I was waiting. The moment I was out of reach there was a click and I rushed back to the phone, lest I lose my turn. It wasn't my turn, of course, just an announcement reminding me that I was still queuing. I was beginning to get cross. It was their fault that the washing machine is broken. Why should I have to pay for the privilege of telling them? When I finally spoke to somebody they said that an engineer would call me back within half an hour. An hour later no one had rung so I started the process all over again. As my fury started to build I could feel my body change. My muscles tightened, face reddened, jaws clenched, hands heated up and heart started beating faster.

Of all the emotions, it is anger with which we associate intense physical feelings. Even something as trivial as trying to sort out a broken washing machine can more than double the heartbeat. A healthy heart can cope with this easily, but for a person with coronary artery disease, it's a different story. The sudden rise in blood pressure caused by rage can occasionally cause fatty deposits inside the wall of the arteries to break off and block the artery. This results in a heart attack or a stroke if blood can't reach the brain. Doing something physical while you are angry, even having a fight, can actually help the body because the vigorous use of muscles causes the big blood vessels feeding the muscles to dilate which in turn begins to lower your blood pressure. Of course a fight might bring you more trouble, but going for a run to 'let off steam' when you feel angry can help you physiologically, which in turn makes you feel better. If you take the alternative, more common approach, which is to sit and fume, then your blood pressure remains high and it's hard to return to a neutral state.

However, this doesn't mean that anger is always 'better out than in'. Repeatedly expressing anger is bad for us, because every time we get into a rage the body gets ready to fight by taking fat from smooth muscle in case extra energy is needed. Any fatty acids not used subsequently cling onto the coronary artery walls, where they contribute to heart disease. Moreover, each time our blood pressure shoots up scar tissue is left by the tiny injuries inflicted on the coronary artery walls, which in turn can contribute to heart disease. The occasional scar is no problem but, theoretically, if this is repeated day after day the harm can start to build up.

Unfortunately this isn't the full extent of the physical problems with anger. When we fly off the handle, the stress hormone cortisol is released; once again this is in case fat needs to be converted into energy at short notice. This is an extremely useful function, but at the same time cortisol diverts energy away from the repair of cells and over a long period of time this could damage the immune system.

If this is what might happen to the body after years of repeatedly losing your temper, then maybe it is better to keep that anger in after all, no matter how furious you feel. In everyday life, however, many believe in the benefits of 'letting off steam' mentioned above; it feels as though bottling up your anger must be bad for you – the sort of thing that gives you an ulcer. It seems not. Calm people get ulcers, as do people who rage, as do people who fume away inside without letting anyone know about it. When it comes to the heart, however, it's not so simple. For years investigators have tried to find out whether it is better to let your anger out or to sit and seethe and in recent research anger has become the focus for studies linking personality and health. There used to be a theory that the possession of a Type A personality (driven, ambitious, energetic and impatient) led to heart problems. In a study carried out at the University of North Carolina in 2000, 13,000 patients were given questionnaires in which they rated their own tendency to get angry. A few years later they were followed up and it was found that although their blood pressure was apparently normal, those who had said they often got angry were three times more likely to have had heart attacks than the others, even when factors like smoking, diabetes and weight had been taken into account. Likewise Mark McDermott from the University of East London found that the people who expressed their anger suffered from more heart disease than those who held back from shouting. Although smoking had more of an impact, anger was making a real difference to people's health. Other studies, however, contradict these findings; some show no link between anger and heart disease, some that the suppression of anger correlates with high blood pressure. The studies are hard to compare because each measures both heart disease and anger in different ways.

In an attempt to get to the bottom of the mystery Giora Keinan from Israel looked not only at how often people get angry, but at the intensity of that anger. She found that in terms of health, the best thing to do is to get really angry, making your case 'clearly and firmly', but to do so only rarely. Keinan suggests the people who only get angry occasionally are likely to be those who are best at finding alternative ways of coping with difficult situations, which reduces the amount of stress they experience and in turn leads to better health. Those who are bottling it up are not using anger effectively to change their situation, but nor are those who always turn to anger as a solution. It seems, therefore, that it's not only whether you let the anger out or not which matters, but how angry you get and how often.

Whether or not you let the anger out can affect your

mind as well as your body. In experiments in Aberdeen Judith Hosie and Alan Milne deliberately made people furious. They showed them film clips from the film *Cry Freedom*, one depicting cruelty to animals and the other a scene in which child protesters are shot dead. Before seeing the first clip one group of people were asked to try not to let their feelings show, while others were told to replace any negative feelings they might have with a happy memory. A third group were simply told to respond spontaneously and naturally to the events they saw in the film. They found that when they were shown the second film clip, the *women* who had suppressed their emotions actually felt more angry than those who had expressed their anger or replaced it with a happy thought. They also felt more angry than the men who suppressed their emotions. This is interesting because it's generally agreed that from a young age girls tend to be taught that anger isn't nice, while boys are allowed to express it. Therefore you would expect the women to be more practised at dealing with their suppressed anger. The researchers speculate that perhaps women are experienced not in suppressing their anger, but in self-distraction or expressing it through tears. When they were asked simply to suppress the feeling, they couldn't. There is also, of course, the problem that anything you are told not to think about remains in your consciousness. There's a classic study known as the white bear experiment where people are asked not to think about a white bear and for them it ends up popping into their minds more often than it does even for people who have been specifically asked to think about a white bear.

twenty-first-century anger

In terms of health at least, it seems that the occasional outburst might be the best way to deal with feelings of anger, but although anger is discussed more these days is it really any more of a problem? 'Going postal' became a common phrase in the United States after various incidents where certain postal workers took revenge on their ex-employers by returning to the offices with a gun and opening fire. Meanwhile we hear stories of road rage, air rage and even trolley rage in supermarkets. Brian Parkinson from Oxford University found that road rage in particular is different from other sorts of anger. There is something special about being sealed in a moving box, unable to communicate properly with other drivers. Brian Parkinson found that arguments between drivers share special features. Firstly, the other driver is usually a stranger, so there is no relationship on which to base your judgements. You also have no idea whether this is the sort of person who is likely to treat you badly deliberately, combined with the fact that the distances and speeds involved in driving make it harder for everyone to make accurate judgements concerning blame. The normal methods of defusing a situation are unavailable when you're in a car. You can't explain your actions, nor use body language easily to show that you meant no harm. You can't apologise in the same way that you would if you accidentally crossed a person's path on a pavement. All you can do is to hold up the palm of your hand, but since this gesture tends to be the driving shorthand for thank you it can make things worse.

These factors are peculiar to cars, but there is an assumption that rage of all kinds is driven by the stresses of modern life. Our expectations of fair treatment might have increased, but after researching anger in depth Mark McDermott from the University of East London believes that there's no evidence to suggest that we are any angrier than we used to be, just that the situations are different. He suggests that perhaps people used to suffer from 'washing-your-clothes-in-the-brook-on-a-rainy-day-rage' but, now that washing machines have taken the frustration out of washing, it has become less of a problem compared with other aspects of modern life, such as traffic jams. We view the world around us, put an interpretation on that world and sometimes our response is anger. Seneca, writing on anger for his brother, listed plenty of aspects of everyday life which made people angry back in the first century AD. He talks of people destroying manuscripts in fury when the writing was too small to read. And his description of an angry man assures us that rage in those days was just as intense as it is today: 'His eyes blaze and sparkle, his whole face is crimson with blood that surges from the lowest depths of the heart, his lips quiver, his teeth are clenched, his hair bristles and stands on end, his breathing is forced and harsh, his joints crack from writhing, he groans and bellows, bursts out into speech with scarcely intelligible words, strikes his hands together continually, and stamps the ground with his feet; his whole body is excited . . . It is an ugly and horrible picture of distorted and swollen frenzy – you cannot tell whether this vice is more execrable or more hideous.'

The big question of course is how you can overcome your

anger. Anger management courses teach people to distract themselves and to look at the problem from the other person's point of view as well as analysing exactly why they feel so angry. Seneca believed that everyday anger with circumstances or objects was foolish because 'objects don't deserve our wrath let alone acknowledge it'. His recommended form of anger management was to accept that each of us has the power to decide whether we've been offended and so if we feel angry we should choose to look at the situation anew. 'The man who has offended you is either stronger or weaker than you: if he's weaker, spare him; if he is stronger, spare yourself.' For those who know themselves to be hot-tempered he advises them not to burden themselves with too many interests, but instead to spend their time reading poetry and looking at green things. He believed it possible to train yourself not to get angry by asking yourself three questions when you go to bed each night: 'What bad habit have you cured today? What fault have you resisted? In what sense are you better?' He wrote that if you put your anger before a judge every day it will soon cease. 'When the light has been removed from sight, and my wife, long aware of my habit, has become silent, I scan the whole of my day and retrace all my deeds and words. I conceal nothing from myself, I omit nothing.'

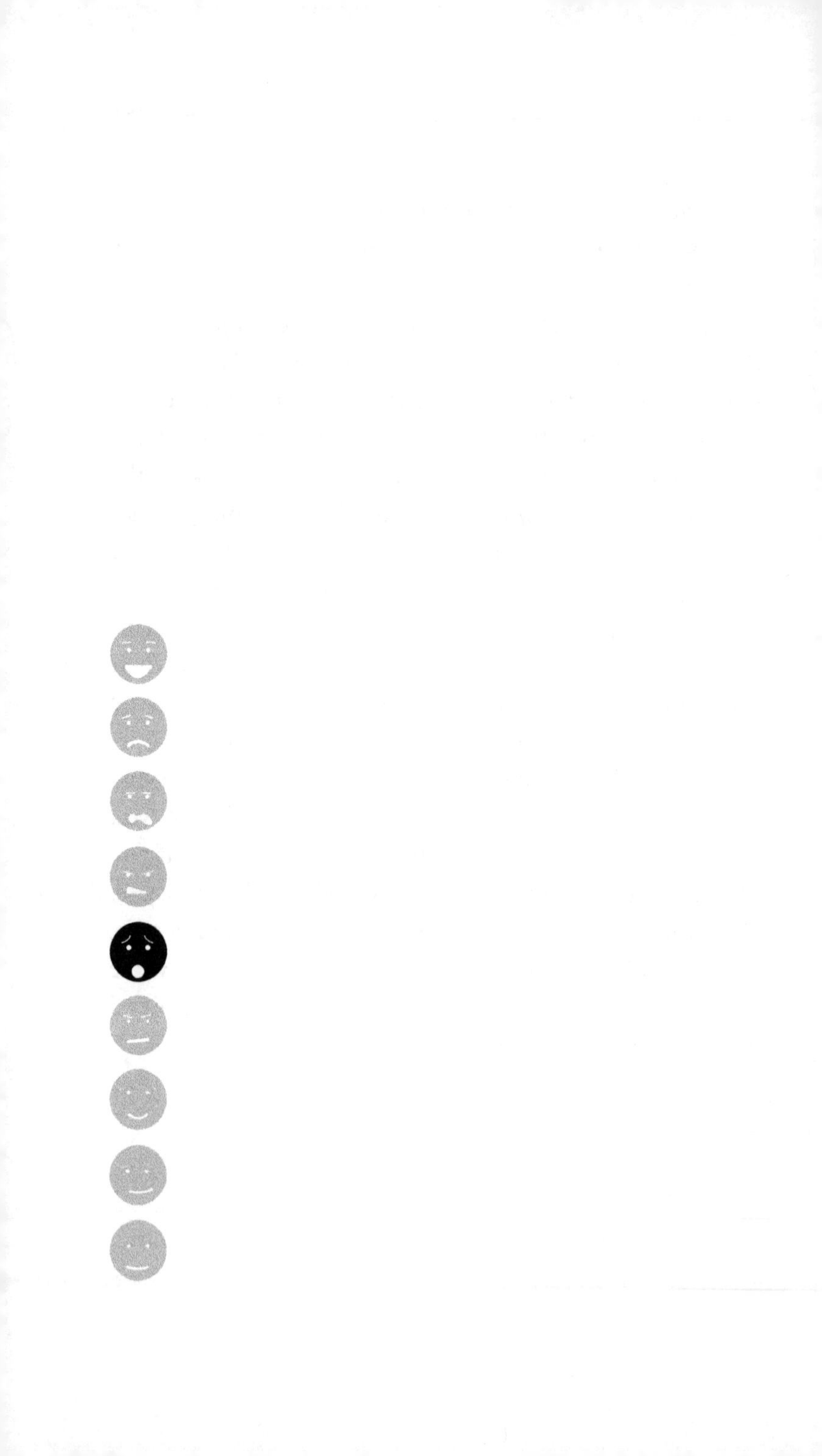

five

fear

One night in June 1984 Captain Eric Moody was all set for a quiet overnight trip from Kuala Lumpur to Perth on British Airways flight BA 009. There was no moon but the sky was clear and flying conditions were smooth. The crew were in a particularly good mood that evening because in addition to their airline meal they had a tray of Malaysian satay. When he'd finished eating, Captain Moody wandered along to the first class cabin. They were flying over Java, but while he was chatting to the purser he was called back to the flight deck. He wondered what was wrong, but wasn't worried, until he climbed the stairs and saw smoke seeping out of vents near the floor. What on earth could be happening? By the time he reached the flight deck he realised things were seriously wrong. There were flames leaping across the windscreen, but worse was to come. Minutes later an engine failed, followed by the other three engines, which stopped one at a time, leaving the huge jet high up in the sky and engineless. A siren began screaming, warning that the cabin pressure had critically fallen. As was all too familiar from the safety demonstration, throughout the plane oxygen masks dropped from plastic compartments and hung in the air. Before the crew could worry about the panic among the passengers they had to get their own masks on. As the Senior First Officer unpacked his, the tube came away from the mask in his hands. At this altitude no one could survive without a mask. Captain Moody had one chance to save his colleague – by losing height fast. Time to talk to the passengers. He put on his calmest, smoothest pilot's voice: 'Good evening, ladies and gentlemen, this is your captain speaking. We have a small problem. All four engines have stopped. We are doing our damnedest to get them going again. I trust you are not in too much distress.'

For the next thirteen minutes he pointed the plane's nose sharply down towards the sea, losing 25,000 feet. Captain Moody should have been consumed with fear, but looking back almost twenty years later he insists he wasn't. 'No one wants to die, but if you don't rationalise it, it gets on top of you. When it happened I'd already had twenty-five years practising keeping my fear under control. It makes you concentrate. I could think very quickly and I could see everything lucidly in front of me. The strange thing was that time went so slowly. At first, when we were heading down for those thirteen minutes I had practicalities to consider, but when I ran out of things to think about I had a moment of real apprehension. I thought we were going to die. It was a silly thought, but what worried me most was the fact that I had £300 in my wallet which would be wasted if we crashed. Then one engine suddenly started up, followed by the others, and I managed to get us to Jakarta airport to land. We were more afraid afterwards. Our biggest fear was that we'd cocked it up and that it would get pinned on us. We got to the hotel at midnight and tried to get pissed, but although we'd had a skinful it made no difference. Our minds were racing. A day-and-a-half later we found out that it wasn't our fault, that we'd flown into a plume of ash from a volcano. When we had to talk through the accident minute-by-minute, going over it all again got a bigger reaction; we felt more fearful then than at the time. After that, euphoria set in. In Singapore a couple of days later we were crossing a ten-lane highway and my friend grabbed me when we were four lanes across, wanting to know why I wasn't looking properly. By then I thought I was invincible.'

What fascinates people about this story is how Captain Moody managed to remain so calm. He thought the plane would almost certainly crash. He was almost certain to die.

Why was there no sense of panic? In fact it's quite common to maintain calm under this sort of extreme pressure. Indeed sometimes people are so serene that they feel detached from the situation, almost as though they're watching a film.

Not many of us will have experienced something quite as dramatic as this, but by adulthood most people have felt extreme fear at some time. On the one occasion when I believed I might be about to die, the fear for my life was accompanied by a calm sensation of intense concentration. It felt as though my brain were working at triple-speed. I was eighteen and was driving back from an interview for a place to study psychology at university, when my car went out of control on a hill, swerving to and fro across the dual carriageway in wider and wider arcs. It was inevitable that the car would eventually crash into the trees in the central reservation, but meanwhile I had time to experience all sorts of emotions other than fear. One was the disappointment that if I died I would never know whether I had succeeded in getting into university. I also recall a bitter frustration at the thought that I would never know what would happen in my life. Then I thought about how my parents would cope. And there was one practical thought: was there anything I could do to try to save myself? At this point I wondered how exactly you die in a car crash. How much would it hurt? Would I be crushed to death? Would the car catch fire? Would I be flung out through the windscreen? For the first time I started to feel real terror.

The whole crash could only have taken seconds, but there was time for all these thoughts, which felt weird and slightly unsettling. The reason time appears to slow down in a crisis is that your brain narrows its focus, paying attention only to crucial information which might save you. So while your brain is desperately scanning through your knowledge and memories of past experience for anything that might be relevant, you cease to notice that ache in your shoulders or the fact that you were feeling peckish. It's not just everyday bodily sensations that cease to matter. Your brain ignores the normal cues which help you judge time. So I didn't notice the time on the clock, whether the radio had moved on to a different song or whether any other cars overtook. Without any clues to time passing, a few seconds can seem like an age.

In extreme cases people even start to see in black and white, a phenomenon mentioned by survivors of the Hatfield rail crash. It's a more primitive way of processing your visual surroundings. When the senses are overloaded and the brain is trying to ensure survival, it's vital that any non-essential information is filtered out. Soldiers fighting amidst the noise of explosions and machine gun fire have been known to report that the battlefield around them appeared silent.

As my brain trawled for anything that might be useful it came up with memories of the contents of aeroplane safety announcements. I bent over in the brace position, covered my head and waited for the crash. Then I was probably knocked out. The next thing I remember was waking up as the car slid very gently towards the ditch and I stared at the pedals, certain that one was the brake, but utterly unable to remember which. At the same time I was aware that there was another way of stopping the car. Even in my befuddled state my brain was trying to grasp that memory and as I looked around me I recognised the hand brake. I was lucky because apart from a bang on the head, I wasn't injured, but I'll never know whether it was my brain-on-overdrive that saved me.

how our fears change

When we start out in life our fear system isn't quite this sophisticated, but right from birth there's one thing that scares everyone – loud noises. Even newborn babies jump with fright if there's a sudden bang. This is known as the startle reflex. One of the reasons babies used to be put down to sleep on their fronts was to suppress this reflex, which in turn stopped the baby waking up so often. Discoveries about the link between cot death and babies sleeping on their fronts put an end to this, but swaddling works on the same principle – suppressing the reflex.

In a sense the startle reflex is an extreme form of surprise, but it is related to fear in that a person who is feeling scared or nervous is far more likely to start when shocked; when children play Grandmother's Footsteps they are wound up ready to jump in fright as soon as the person playing the grandmother suddenly spins around. They jump far more severely than if someone facing the other way had just turned around and made eye contact with them. The American anthropologist, Ronald Simons, travelled around the world collecting 'startles' and found no shortage of people who jump out of their skins at any opportunity. They're known as hyperstartlers. He found one woman called Mrs

Gould who jumped so severely that on one occasion she accidentally bit somebody. She was working as a waitress in a late night diner, when a customer came in asking for fresh orange juice. As she returned from the kitchen she was concentrating so hard on carrying the pile of oranges, that when a man scrubbing the floor said 'hello' she screamed, threw the oranges into the air and dropped to her knees, somehow managing to bite the poor man's thumb as she fell. Curiously Mrs Gould said although she jumps a lot, she's not a nervous or fearful person. In fact she's so blasé about safety that her family worry about the fact that she leaves her front door unlocked and is happy to wander around cities on her own late at night. Although in the West people who startle easily might be mocked, in some cultures, for example, certain Malaysian communities, the same behaviour is revered.

Simons did his own startling experiments in the lab. He warned people that at some point he'd let off a pistol and then filmed their reactions. To enable him to identify the components of a genuine startle, he would ask people to either try to fake a startle when there was no gunshot or to disguise a startle when there was, both of which are hard to do. He found that even with a countdown from ten to indicate exactly when the pistol would go off hyperstartlers still jumped out of their skins.

These people were paid for their time and had the option to leave the study if they so desired but two centuries earlier an experiment was conducted along rather more sinister lines. Thomas Day, a follower of the philosopher Jean-Jacques Rousseau, selected a pretty twelve-year-old orphan called Sabrina and set out to mould her into the perfect wife. He experimented to see whether he could make her immune to fear by finding different ways to frighten her such as letting off a pistol next to her ear and on one occasion even setting fire to her petticoats. Unsurprisingly this didn't work and, fortunately for the girl perhaps, Thomas Day later died after falling off his horse and she married somebody else.

The startle reflex is the only form of fear shown by newborn babies and not every startle involves fear. Surprisingly babies might not experience real fear until they're about seven to eight months old, although there are some reports of observations of fear a couple of months earlier. For example, when Jake, the son of a friend of mine, was five months old he would look alarmed and cry whenever he heard a hairdryer or vacuum cleaner. By the time he was eight months he was afraid more often. Green Goddesses, the army's fire engines, were out on the streets during a firefighters' strike and their sirens would frighten him so much that his parents had to lift him out of his pushchair to comfort him. When he was younger and was carried close to his parents' chests he seemed less afraid. From the point of view of survival a baby's sensitivity to fear needs to increase as soon as they have a little more independence from their parents. By crying out Jake ensured that his parents kept him safe from something large, noisy and unfamiliar and therefore potentially dangerous. Babies are not afraid of strangers from birth, but become particularly wary of them at about the same time as they learn to crawl. Intriguingly, whatever age babies are when they learn to crawl, within the next few weeks they become afraid of heights for the first time. Again, this keeps them safe. As well as a fear of strangers and heights babies seem to be predisposed to develop fears of three particular conditions: when they are left alone, approached quickly or when a situation is in some way strange.

kwolls and kwokkas

A ten-year-old girl approaches a wooden box in the corner of the primary school classroom in Uckfield, East Sussex. She glances at me nervously and gingerly puts her hand into the dark hole cut in the front of the box. 'It's soft,' she says, looking relieved, 'and it's got a long tail. Can I see it?' She's taking part in an experiment devised by Andy Field, a psychologist from Sussex University. The idea is to see whether simply telling children negative stories about animals can make the children fear them. They've picked three obscure Australian marsupials in the hope that these children from a small English country town won't have come across them, but they are genuine species – the kwoll, the cuscus and the kwokka. The information the children are given about the animals, however, is not so genuine. One is described as vicious and unfriendly, while another is cuddly and sweet. The experiment is a success and the children do become afraid of whichever animal about which they've heard negative information.

As with babies, children's fears go through different stages. Before the age of seven they might be afraid of ghosts, the dark, monsters and animals. By ten or twelve they start to fear events that, although real, are very rare, such as murder or nuclear war. When I asked a group of twelve-year-old children what scared them they brought up subjects like terrorist bombings and paedophiles. At their age I remember being afraid that if I ever went to the United States I might be framed for a murder and put to death in the electric chair. I can only assume that I'd seen a film where this sequence of events occurred.

Researchers from the Netherlands found that parents are often unaware of just how much their children are worried by these sorts of things. They also found children are surprisingly good at developing coping strategies for dealing with their fears, particularly at night. The children reported hiding under the covers, switching on the light, deliberately distracting themselves by reading or trying hard to think about something nice.

By adolescence I'd realised that if I were ever to visit the States, I probably wouldn't be framed for murder or executed. With the exception of lasting phobias, unrealistic fears like this tend to disappear during the teens. However, if children are very shy, they're more likely to remain afraid and occasionally they develop generalised anxiety disorder as an adult, where they become scared of virtually everything.

Once teenagers have ceased to fear that they will come to some sort of physical harm, they have a whole new world to fear. Murderers and monsters no longer terrorise them. Instead they're frightened of being rejected or looking foolish.

By adulthood people have developed a variety of ways

of coping with their fears and can, to an extent, distract themselves if they do feel scared. Sometimes these coping strategies are so effective that it's not until they're disrupted that they remember that a situation is frightening. It was found, for example, that if music was played in a blood donation centre, people were less likely to come back to donate a second time. The music was intended to relax them but it seemed that a lot of people coped with lying on a strange bed while a needle siphoned blood out of their arm by simply not thinking about it. They distracted themselves by deliberately thinking about other things but the music seemed to bring their awareness back into the room, whereupon they realised they were scared and this deterred them from coming again.

In more extreme situations coping mechanisms aren't always enough and feelings of fear can even hinder survival. Throughout history there are stories of soldiers cut down by the enemy after becoming paralysed with fear. The Byzantine Emperor, Theophilus, according to the French sixteenth-century essayist, Michel de Montaigne, was so overtaken by fear in one battle that he was unable to move. In the end one of his commanders threatened to kill him if he didn't proceed. Montaigne wrote that this uncertainty over how one might respond to fear made fear itself the thing he was most afraid of; 'Sometimes fear puts wings on our heels; at others it hobbles us and nails our feet to the ground.'

The First World War provided an unwelcome opportunity for many to find out how they would respond in the face of daily terror. At the extremes, some developed a foolhardy bravery which sometimes put them into more danger, while others were shot for cowardice when in fact they were traumatised. In evidence to the 1922 War Office inquiry into shellshock a medical officer described seeing a man on the beach at Gallipoli suddenly become possessed by fear: 'He saw the whole beach covered with jewelled spiders of enormous size. They did not know what to do with the man so put him on one of the boats, and as the barges came up with the wounded he thought he saw his wife and child on a barge, cut in pieces.' The more barbaric treatments for shellshock included the application of electric shocks to the voicebox, cigarette burns to the tongue or agonising lumbar punctures. Public attitudes at the time weren't always the most sympathetic, but a letter to *The Times* in 1920 from a member of the public called C. M. Wilson summed it up. Although he didn't claim to have any specialist knowledge in this area – and notwithstanding his lack of sympathy towards those who were traumatised in the trenches – his observations do get to the nub of fear. 'First there was the man who did not feel danger; secondly the man who was afraid and did not show it; thirdly the man who was afraid, showed it and did his job; fourthly the man who was afraid, showed it and shirked.'

None of us can know how we will react in a terrifying situation. Perhaps this is why fear is so dreaded. If we're lucky it might provide us with a strength and speed of thought that we could never have imagined, as it did with Captain Moody's colleague. He succeeded in using his dexterity and intelligence to mend his broken oxygen mask whilst being aware that he was slowly suffocating. Unfortunately until it happens we can't be certain whether we might be a cool Captain Moody or the poor man who saw jewelled spiders.

the fearful brain

One of the people who pioneered the change in attitudes towards research on emotion was Professor Joseph LeDoux from New York University. Rather than trying to pin down the nature of the conscious experience of feelings, he set out to discover how one part of the brain can signal a feeling both to other parts of the brain and to the rest of the body.

The brain's emotional system is thought to have evolved before the parts of the brain that deal with complex thought processes. Fear was probably the first emotion to have evolved, so essential to survival that some say it was probably experienced by early vertebrates living four to five million years ago.

Considering the influence that emotions can have on our lives, surprisingly few areas of the brain deal with them. Picture a nice diagram of the brain divided into sections, with one labelled 'emotion' and that area subdivided into smaller sections, labelled 'fear', 'guilt', 'joy' and so on. Sadly that's not how it is, although it would make life easier for neuroscientists if it were. The brain does deal with each emotion in a different way, but rather than there being separate areas for each, it seems that combinations of several areas are used to a greater or lesser extent. Although neuroscientists are just beginning to work them out, each emotion appears to have its own pattern. As I've mentioned before, one particular part of the brain – the amygdala – is often seen as the key to the emotions and to fear in particular.

You're walking along the street late at night in an unfamiliar area, and you're hypersensitive to anything that goes on around you. As you pass a driveway you see someone out of the corner of your eye – a man crouched down next to a rubbish bin. You jump, but the brain's response is more complex than the well-known fight or flight mechanism. According to LeDoux's theory, the fear you're feeling now has come about via the quick and dirty route into your brain. In an instant your brain has registered a threat and deep inside, the amygdala has received the news and sent out a message to tell your body to respond. Your heart starts beating faster, your palms sweat, goosebumps come up on your arms and your hairs stand on end. Until this moment you might not even have consciously realised that you are afraid.

Seconds later you notice that the dark shape is just a dustbin bag. The brain's second route to fear has kicked in. Alerted to possible danger, you've assessed the situation, focusing all your attention on the dark shape until you've worked out what it is and used your previous knowledge to judge that the bin bag isn't likely to be a threat. You relax.

The initial response is like a car alarm that's been set a little too sensitively and starts wailing whenever a big lorry drives past. With cars it can be extremely irritating, but with humans the system needs to be on the sensitive side. It's better to jump momentarily when you see a bin bag than to fail to notice someone crouching in wait who might be out to do you harm.

Cleverly, even with the quick and dirty route the brain still takes account of the context. So provided a person is not phobic about snakes, they won't recoil in terror when they see a snake in a tank at the reptile house. Even before they've had time to reason through whether the snake could hurt them, the brain is aware that they're in a zoo which isn't generally dangerous, so it's as though the brain is on a different sensitivity setting. If, on the other hand, the person woke up in a tent in the rainforest to see the same poisonous snake sliding along the sleeping bag towards the pillow, the reaction would be rather different. As well as context, memory and imagination can influence the amygdala. So if you're driving along a badly-lit road through some woods thinking about the famously frightening film *The Blair Witch Project* then, not surprisingly, you'll be more jumpy. This is why sitting in a deserted churchyard at night telling ghost stories is a particularly effective way to terrify yourself as a teenager. You can prime your fear like a spring, ready to be released at the slightest noise.

The ingenious thing about fear circuits is that once we've learnt how to avoid a particular danger the amygdala stops being involved. Every day we come across potential dangers that no longer scare us because we know how to deal with them. It would be hard on the nerves if we froze with fear every time we reached a main road and caught sight of a fast car. Instead, once we've learnt how to cross the road safely, the amygdala keeps out of it – unless something unexpected happens, like a car driving on the wrong side of the road just as we're about to step off the kerb. Then the amygdala is ready to set off those alarms again.

the fearful body

It's 1953. A man is sitting at a table with electrodes attached to the back of his hand, connecting him to a machine. He's been warned that he might be given mild electric shocks, but he's pleased to be playing his part in the progress of science by participating in an experiment and knows the shocks won't be powerful. The future depends on science after all. The experiment has just begun when there's a sudden bang and sparks start showering out of the machine. The scientist conducting the experiment is visibly panicked, pressing buttons at random and hammering on the machine, but it won't stop. Finally he gets to the socket and rips the plug out. The man taking part in the experiment is unhooked and taken to a second room where another scientist is waiting to begin the next part of the study. He rudely orders him to sit down and, whilst attaching him to a lie detector, berates him for being late. The human guinea pig, we can assume, is by now regretting ever having volunteered for the research in the first place.

Of course the whole thing was a set-up. The experimenter Albert Ax was the first person to measure how the body responds to fear and anger. To do this properly he needed to provoke genuine emotions so, in exchange for a three-dollar fee, he terrified the poor participants and then confronted them with an obnoxious experimenter. Some of the people were so frightened that they begged the experimenter to disconnect them from the machine and at least one man prepared himself for the fact that he might well die. Another, furious with the rude experimenter, said afterwards, in wonderful fifties phraseology, 'Say, what goes on here? I was just about to punch that character on the nose.'

Back in the 1950s the experimental conditions weren't ideal; apparently the polygraph made a blip every time the lift stopped at that floor. Although Ax was careful to preclude those with high blood pressure from the study, this sort of experiment would never get past an ethics committee these days. However, he was able to confirm for the first time under controlled conditions that fear has definite effects on the body, increasing heart rate and sweating, as well as speeding up breathing and causing sudden rises in muscle tension. Today the basic fight or flight mechanism is well-known. Adrenaline levels shoot up, your heart pounds faster, skin temperature drops, pupils dilate, tolerance to pain increases and energy is diverted away from everyday tasks such as repairing cells or digesting food and is concentrated instead on your muscles, particularly in the legs, so as to provide the means for escape or battle.

Any number of people can tell their anecdotal reports to researchers, but the only way to confirm the actual physiological effects was in experiments like that of Ax. The key was to induce enough fear to produce the genuine emotion, but not so much as to traumatise the volunteers, or worse, kill them. In an ideal world you would wire somebody up before they went through a real-life traumatic experience but the trouble is, you never quite know when something like that is going to happen.

scared to death

On 17th January 1994, Steph was woken suddenly in the night by the feeling of someone shaking her bed. While out in Los Angeles working as an entertainment reporter for a British TV company, she was living on the ground floor of a Spanish-style 1950s duplex on Pico Boulevard, a wide street lined with palm trees in west Los Angeles. But her trembling bed had nothing to do with anyone human. 'I heard this incredible rumbling sound and was thrown out of bed. The wooden floors were undulating as though they were fluid, like rippling waves. You could feel the earth going up and down underneath and things were crashing and breaking. It was petrifying. In the pitch black the very structure of the house seemed to be cracking. I was frightened the ceiling would fall in on top of me and so for some strange reason I grabbed the duvet off the bed and put it over my head thinking that it would in some way make me safe. I was being thrown back again and again, banging against the wall. I've never been so frightened. After about thirty seconds I realised it was an earthquake and I was even more petrified, but there was nothing I could do.'

Steph was unhurt and hasn't been put off returning to work in a city so close to a fault line. Had it been possible to measure her brain's response while she cowered under the duvet, it would have provided the perfect opportunity to discover more about fear. However, the Los Angeles earthquake did reveal something else quite extraordinary. More than one hundred people died, but not because they were all hit by falling masonry or buried under debris. When the coroner's records for that night were compared with each night the week before and the same night in each of the previous three years, they found that five times as many people died from cardiac arrest during the earthquake than on a typical night. Witnesses saw people clutch their chests and complain of pain, before collapsing to the floor and dying.

In the minds of the people of Los Angeles one idea has long been lurking: that one day an earthquake so big will hit the state that the entire coast will be devastated. It's no wonder that at 4.31 that morning, some people thought their end had come. What's more, the week after the earthquake there were fewer heart attacks than usual, suggesting that the people with a heart weakness who might have had a heart attack the following week had already died during the earthquake.

The phrase 'scared to death' seems to have some truth in it. The fight or flight response, which usually works so well, giving us the focus and energy we need to deal with a situation, can backfire by stopping the heart completely. A burst of adrenaline makes the blood vessels constrict with the sensible aim of preventing blood loss if a person is attacked. The problem is that simultaneously the cells in the heart fill up with calcium and occasionally this can force the heart to contract so strongly that it never relaxes again and stops altogether. This is more likely to occur in someone who already has heart disease, but occasionally a healthy person can be affected in this way. The irony is that the very emotion that's evolved to keep us safe, which developed in creatures long before any other, can be the one that kills.

Fear can have useful physiological effects, however, and not just when we need to run away from a bear. A little fear before undergoing surgery can help the body to deal with the physical trauma of an operation. In a UK hospital Anne Manyande, a researcher from Thames Valley University, found that although patients who were given relaxation training felt less anxious prior to surgery, their bodies responded less positively. Both during and after surgery they experienced surges in the stress hormones adrenaline and cortisol, which could impair the immune system. These were the very patients who were so relaxed beforehand that they had lower heart rates and blood pressure than the other patients. This suggests that a degree of fear before surgery somehow prepares the body for what it's about to undergo. If you're too relaxed the body can respond over-vigorously to the shock of surgery, producing large amounts of the stress hormones. The solution to this isn't for doctors to terrify their patients before surgery, but to talk through the procedure and to explain the nature of the pain they are likely to experience later. This gives them the chance to prepare themselves mentally for what the body is about to go through.

the smell of fear

There may be one more way in which the body responds when we're afraid. Is it possible that we have a warning system which lets people smell our fear?

Most of the houses in the windy, coastal town of Tangalle in Sri Lanka consist of concrete blocks with metal spikes protruding from their flat roofs, forever awaiting the next storey, which never arrives. The pavements, where they exist, are dusty and cracked, with rubbish piled up at the corners. A hundred yards along the road out of town, and sights are rather different. There's a sweeping curve of honey-coloured sand sloping steeply down to a sea full of breakers which attracts tourists from across the world. Skeletal catamarans painted in yellow, red and blue are hauled onto the beach by fishermen. Little boys make chaotic backward somer-saults off the stone quay into the water. Behind the quay there's a larger sandy stretch protecting a lagoon where clusters of restaurants on stilts sell the local arak to people like me and my boyfriend, who were travelling around Sri Lanka ten years ago. One day we set off to find out what was around the next headland. As we walked along yet another canary yellow beach, small, black crabs scattered beneath our feet, slipping away into round holes in the sand. At the far end of the bay there was a grassy headland which promised a good view back across the beach to the jungle beyond, but the moment we reached the headland we sud-denly felt chilly. The paradise beach atmosphere had gone. There was a distinct smell in the air. A couple of large bones were lying on the grass. Maybe shark bones brought in by the tide, we thought. But moments later we reached a pit where raucous crows were picking over dozens of bones and weighty skulls. At the tip of the promontory there was a small roofless stone hut, open on one side. A horizontal timber pole was stretched across the hut, suspending four rope loops. We realised this must be a primitive cattle abattoir, situated away from the town in what happened to be a spot so beautiful that no doubt visitors had walked out here before. What was extraordinary was the smell. The remains of rotting flesh clung to the bones of the slaughtered cows and there were bloody stains on the ground, but there was something else as well – the smell of bovine fear perhaps.

It has long been known that some animals do release a specific scent when they're afraid. Frightened fish let off pheromones which serve as an alarm signal to other fish. Even dead fish can leave warnings for others. When a fathead minnow is eaten by a pike it cunningly leaves a scent behind on that pike which induces fear in minnows who have never before seen a pike. Bees, lice and earthworms have similar mechanisms and ants release warning pheromones which alert other ants either to disperse or to get ready to defend the colony.

There is of course always a danger in extrapolating from animal behaviour to human behaviour. You can choose an animal to demonstrate just about any view you happen to have on humans, but Professor Karl Grammar from the University of Vienna wanted to know whether humans might emit fear pheromones too. In the process of finding out he ended up with a freezer filled with armpit swabs.

When the volunteers arrived at his lab they were sent to the bathroom to wash under their arms and change into clean, white T-shirts. Then they sat watching the film *Candyman* with a cotton pad wedged underneath their armpit. This horror film tells the story of a student investigating a local myth about a black slave who died after a racist mob cut his hand off, covered him in honey and set a swarm of bees on him. The legend has it that if you look in the mirror and say 'Candyman' five times he returns to murder people using the hook which replaces his hand. Naturally the student decides to establish whether there's any truth in the story. You can probably guess what happens but, suffice to say, Professor Grammar succeeded in his aim of scaring the volunteers who watched the film. The armpit pads were collected and put into the freezer for preservation. A few days later the volunteers came back and this time they watched a video of a train journey filmed from the driver's cab. Since the train didn't crash this film wasn't in the least bit frightening.

Having to watch the whole of *Candyman* sounds bad enough, but volunteering for the next part of the experiment was probably worse. People had to sniff the armpit pads from each film and rate them for intensity of smell, pleasantness and whether each pad reminded them of sex, aggression or fear. They found that the people watching the horror film did have stronger, more unpleasant-smelling sweat, but this wasn't due to the levels of the stress hormone, cortisol. Curiously the smell was judged to be aggressive rather than fearful. It is too early to say, but maybe frightened humans release a smell of aggression with the aim of deterring opponents.

when the fear system goes wrong

I've already said that our fear system works like an alarm that's set to alert us to danger, with the inevitable false alarm from time to time. However, the system can go wrong, becoming oversensitive to potential dangers, particularly after a traumatic experience. People with post-traumatic stress disorder (PTSD) scare much more easily. They have nightmares and flashbacks, where adrenaline shoots up to levels almost as high as they were during the original incident. Situations with any similarity to the circumstances of the trauma become so terrifying that people avoid them, even if it means missing out on things they enjoy.

It has to be remembered that it is not only what actually happened that scares people; it's what might have happened. When I visited a residential home to speak to Second World War veterans who were spending time there in order to help with their PTSD, I met a man who told me his extraordinary story. He had been standing in the turret of a tank with his friend when the tank was hit. As he turned to his friend he saw him decapitated. He himself was uninjured and survived the rest of the war, but ever since he has been haunted by the idea that something so appalling could easily have happened to him as well. More than fifty years out of danger and the nightmares still persist. His fear ceased to have a protective function long ago. As a retired man living in a quiet Dorset village, there's no chance of him being in a tank under fire ever again. The fear system, which should be there to protect him, has gone wrong.

Due to the variety of brain systems which feed into each

other to produce fear, these faults can occur in a number of ways. If the amygdala, the walnut-shaped organ which I mentioned earlier, becomes oversensitive the slightest happening will provoke fear. The mind seems to over-generalise after trauma, so a slamming door might remind a traumatised person of a gunshot. Alternatively the amygdala might detect threats accurately, but provoke over-extreme responses, prompting a person to react with terror at a situation which is a little strange, rather than dangerous. For example, if the phone were to ring a couple of times and then stop, most people would be puzzled at most, but for someone who was once attacked in their home in the past, this might induce terror.

There is also a third way in which the amygdala could malfunction. It might both detect and send out reactions normally, but the system that overrides fear – telling us that a danger has passed or that in this type of situation you are safe – might not kick in, leaving a person feeling terrified. The fact that the system can go wrong in all these different ways shows just how complex the fear process is, with different parts of the brain working together in an attempt to produce the appropriate emotion at the appropriate moment.

The organisation of the brain could shed some light on the variation in the effectiveness of different therapies in dealing with fear. Cognitive behavioural therapy tends to work better with phobias than does more in-depth psychoanalytic psychotherapy where memories and thoughts are explored in detail, enabling the client to gain a deep understanding of the problem. The area dealing with working memory and thoughts is called the lateral prefrontal cortex, but there aren't many links between this area of the brain and the amygdala, where fear is centred. In cognitive behavioural therapy, however, the client learns to replace unhelpful thoughts with new ones, work which is likely to take place in regions further back in the brain which do have more links to the amygdala. This is only a hypothesis, but it would be fascinating if the layout of the brain could explain why certain problems respond better to particular types of therapies. Unfortunately this idea doesn't explain why a therapy that works for one person doesn't always work for another.

There are good evolutionary reasons for avoiding dangers you have experienced before, but the big question with post-traumatic stress disorder is why some people become traumatised to an extent that goes far beyond self-protection and that can ruin their lives. One intriguing theory involves beta-endorphins. These are the body's natural painkillers. When a person is injured endorphins are released to help with the pain. People who have been severely traumatised seem to have higher levels of endorphins than everyone else. It's not known whether these levels were high before they experienced the trauma. This is tricky to research because you can't measure endorphin levels with a blood test, but what you can do is to scan the brain and see where neurotransmitters are activated. The problem is that to find out what happens to endorphin levels during a terrifying event, you would have to scan somebody's brain while you traumatised them; not something for which you would find many volunteers. This research is in its infancy, but what we need to know is whether the raised endorphin levels are causing the symptoms of post-traumatic stress disorder, such as nightmares or flashbacks, or whether those symptoms are raising the levels by causing the body to re-experience the trauma. It is possible that these neurotransmitters and stress hormones rise during a traumatic incident as part of a defence mechanism, but that when the event is over for some reason the levels don't always return to normal. Perhaps the memory of the event is so profound and intrusive that a high level of physical activity in the brain continues, almost as though the event were still happening. The system that is there to protect us ends up taking control.

the twilight world of intensive care

In some senses it is logical that PTSD might follow a traumatic experience involving an attack. However, extraordinary research has found that in one particular situation PTSD can even follow incidents in which a person was not under attack but cared for.

You're lying on a bed in a bright, white room filled with equipment. You cannot move. There are machines beeping, with green traces skimming across electronic screens. In front of you there's a gleaming white console with rows of switches and yet more screens, like the controls on a spaceship. Black squeeze boxes move up and down of their own accord inside glass tubes. Men and women in white coats rush past. You try to speak but nobody hears. You're voiceless. They give you injections and you can't stop them. You can't move. They tell you it will be fine. What are they injecting into you? Are you part of an experiment? Nothing looks familiar. You don't remember coming here. Have they brought you here to kill you?

This is not a scene from a sci-fi film, but an experience that's not uncommon amongst patients in intensive care. While their relatives sit beside the bed, assuming they're resting peacefully, made comfortable by powerful drugs, a lot of patients are in fact experiencing hallucinations.

Extraordinarily, as many as 30% of people who have a stay in intensive care suffer from post-traumatic stress disorder (PTSD) afterwards. This is a high percentage, considering that only 3.5% of victims of assault end up with PTSD. At Whiston Hospital in Merseyside, Christina Jones has spent the past ten years attempting to ascertain the reasons why a stay in intensive care can cause such problems. Many patients remember very little about their stay because they were brought in unconscious and then sedated. Surprisingly the distress that patients report afterwards does not tend to relate to the fact that they almost died, nor to the physical procedures they underwent in hospital. In fact patients often find it hard to believe how seriously ill they were and become frustrated both with their slow recovery and the extent to which their relatives fuss over them. Their families of course remember every moment of the days or weeks spent witnessing their condition in hospital. At the Whiston Hospital nurses keep diaries for the patients explaining exactly what happened to them in intensive care, complete with Polaroid photos. To show someone a photograph of themselves looking this ill may sound cruel, but it does seem to help patients to accept the gradual pace of recovery.

Gillian spent four weeks in intensive care after she was rushed to Whiston Hospital with breathing problems. She turned out to be seriously ill with pneumonia. 'When I first woke up I thought I was in New York and that I was Michael Jackson's sister, but I was white. I went to have my teeth done red, white and blue and one of the girls doing my teeth was jealous of me for being Michael Jackson's sister. I thought the nurses' station was a space ship. All I could see were these little people with antennas, like something out of the Teletubbies. They were experimenting on me and there were all these chickens around that they were basting. It all sounds really strange, but it felt so real. When I woke up later and saw one of the nurses I'd seen experimenting on me I was terrified. I didn't want her coming near me.'

These hallucinations are known as 'ICU psychosis' and although the condition was noted by intensive care doctors back in the 1950s, it is rarely discussed outside medical circles. These hallucinations are different from the near-death experiences which some patients describe and even find comforting. Aliens often seem to play a part, as does the idea that the doctors want to kill them or use them as experimental guinea pigs. The hallucinations can be terrifying and at the time they feel more vivid than dreams, leaving people with distinct, but frightening memories. One study in Germany found that some patients were still suffering from PTSD ten years after a stay in intensive care. At many hospitals the problem is never picked up because patients do not return to intensive care after recovery, but to a specialist for their particular illness. That specialist might only see the occasional person who's been in intensive care so they are unlikely to start asking whether a patient is having flashbacks. For their part, patients are often reluctant to mention the strange things they remember seeing while they were in hospital.

The reason for the hallucinations is two-fold: they can be caused by the delirium connected with serious illness, but hallucinations are also caused by treatment with some drugs and withdrawal from others. These drugs not only reduce a patient's pain, but crucially for the staff, they keep the patient still. A sedated patient is easier to treat than one who is trying to pull their tubes out in panic. The trouble is that sedation is complex and the ideal levels aren't known. Would it make a difference if patients were slightly more awake or slightly more asleep? Christina Jones at Whiston Hospital believes we are chemically restraining patients without understanding the possible side-effects. (In some southern European countries including Italy and Spain patients are not routinely sedated. Instead they are physically restrained to ensure that the equipment stays in place. This sounds barbaric, but if it prevents hallucinations, it might lead to less PTSD. A study is currently underway to see whether this is the case.)

the memory of fear

On 6th March 1987 an ex-sailor sat at home watching television, waiting for the news to come on, unaware of the impact that this particular news bulletin was to have on his life. The headlines told of a ferry which had capsized just off Zeebrugge – the *Herald of Free Enterprise*. The pictures showed the ship lying on its side, its vast red hull parallel to the sky. Just ninety seconds after the roll-on roll-off ferry had left port with its bow doors still open, enough water had flooded the bows for the entire ship to capsize, killing 193 people, many of whom died from being thrown around inside the boat. The man sitting at home watching the news knew no one on the ferry, nor had he even worked a ship since leaving the navy four years earlier. That wasn't the

problem. The problem was that the stress of hearing survivors of the ferry disaster recount their experiences brought it all back to him. The vast ship, defeated and drowning. The screams. The flames. The terror. During the Falklands War he was serving on a guided missile destroyer, HMS *Coventry*, when it was hit by two 1000lb bombs just off Pebble Island. Water had poured in straight away and the order had been given to abandon ship. Swimming through the freezing water he had somehow reached a life-raft and was dragged on board to join the other men, but every wave brought the raft surging back towards the burning ship, no matter how hard they paddled. The ship was leaning over towards them and they knew that if it capsized their raft would be crushed. Two helicopters veered towards them and used their downdraughts to blow them away from the flames. As soon as they had made enough distance from the ship, one of the helicopter pilots hovered daringly low, aware that the ship might explode at any moment. A line wavered down towards them and one at a time the men were picked out of the raft and winched up into the helicopter. At last he was safely inside the helicopter, but for nineteen other men still on the ship there was no escape as minutes later it capsized and began to sink slowly beneath the water.

A year later he left the navy and began life as a civilian, putting the events of the Falklands behind him. He remembered the day well and those who died remained in his thoughts, but he tried not to dwell on the details. Civilian life was certainly different, but he and his wife saw a lot more of each other and things were working out well. Seeing the pictures of Zeebrugge changed all that. He couldn't bear the idea that other people were going through the same appalling experience, but with such a high death toll. That night and every night for months afterwards he dreamt that he was back on HMS *Coventry.* He remembered the men's screams as the boat tipped. He remembered watching his friends desperately trying to heave themselves out of the freezing water and up over the edge of the raft. He began to dread sleeping, anticipating the terrifying nightmares sleep would bring, but the daytime was no better; at work he couldn't concentrate and sometimes he just found himself staring at the wall for ten minutes at a time. On the way home he would sit on the bus and suddenly he would see not the backs of heads in front of him, but the burnt faces of men clinging onto the raft. Then he would see the end of the road and realise he had missed his stop yet again. Passengers would stare as he sobbed. The ex-sailor was suffering from post-traumatic stress disorder, but his symptoms had been delayed for years.

This is not at all uncommon. Often symptoms occur after some other stressful event later in life. At the end of the twentieth century, psychiatrists were seeing men who thought they had coped with the trauma they had experienced while fighting back in the Second World War, but after retirement or the death of their wife the nightmares and flashbacks to the war began. In terms of the brain it's rather curious that stress in later life can revive frightening memories from years before. The content of memories is laid down in the part of the brain called the hippocampus. Meanwhile it's thought that feelings about the memory are stored in the amygdala, separately from thoughts about the content. In infants the amygdala matures before the hippocampus which could account for our inability to recall memories from the first couple of years of life. We can't recall these memories because they were never laid down. In one experiment on the memory of fear twelve-month-old babies went for their routine injections. They all cried when they felt the needle pierce their skin. Two months later they returned for further injections from the same doctor, but they did not cry when they saw the doctor approaching.

Even at that age they didn't remember either that this doctor had hurt them previously or even that a person holding a syringe near your arm is someone to fear.

It's thought that stress can impair the ability of the hippocampus to store memories, but meanwhile the amygdala still forms the emotional part of the memory – how the event actually felt as opposed to an intellectual description of what it felt like. So while the hippocampus is remembering details of the street where you were attacked, the amygdala is remembering the surge of fear you felt. The idea is that the emotional memory can remain even if you can't remember a very stressful event. These memories of fear are usually kept under control, but during stressful times this might become harder and somehow they're released. Professor LeDoux speculates that for the first time this might provide a biological explanation for the psychoanalytic idea that traumatic experiences in infancy might affect your whole emotional life. Although we cannot remember the content of early experiences, perhaps their emotional flavour stays with us.

Professor LeDoux hopes to find a way to use the brain's chemistry to reduce the symptoms of PTSD. In order to recall a memory the brain makes new proteins. If someone is plagued by horrific memories, in theory – and at the moment it is only a theory – you could stop the production of those proteins, thus preventing the retrieval of that memory and the ensuing distress. The problem is that drugs can't be injected directly into the amygdala and if a drug were injected directly into the bloodstream protein synthesis would shut down all over the body, causing chaos. An alternative approach is to find a gene which blocks just the right protein in just the right place, but at the moment no one knows how to do this. If this were possible in the future it would raise the question of whether you would want your memories expunged when they form a part of your personality. People with amnesia complain that their entire identities are lost. It might be useful to remove the memory of a one-off traumatic incident, but if a person's whole life has been built around their memories of a particular time, then the loss of those recollections could be devastating. Take Primo Levi, who wrote a series of extraordinarily powerful books about his time in a concentration camp during the Second World War. How much did his experiences become part of his identity? His books have had a huge influence, but he wrote them long after the war. Had his memories been disposed of, those books wouldn't exist. Despite the enormous pain that his memories caused him, leading to severe depression and maybe even to a death that some say was suicide, he was distressed when he felt his memories fading with age. He told a friend that he was deliberately rereading his own books in order to remember what happened. Far from feeling relief that the recollections were becoming fainter, he didn't want to forget.

the mind's ability to reduce fear

At a wedding on a summer day in 2003 I was starting on my roast guinea-fowl in a cream sauce when the person next to me said, 'You see that man on the other side of the table – apparently he frightens mice for a living.' Intrigued

I went to talk to him and found it wasn't as heartless as it sounded. His name was William Falls and he runs a laboratory at the University of Vermont. His approach to the problem of fear is slightly different from that of Professor LeDoux. Instead of trying to block the proteins that enable us to recall a traumatic memory, Falls is examining the way we naturally deal with fear by inhibiting it. As I mentioned before, when we're afraid we employ strategies, often without even realising it, that calm us down. We might deliberately think about something else or look around for familiar things that comfort us. Using our minds we do have incredible power over our fears.

Falls' theory is that people with post-traumatic stress disorder or extreme anxiety are for some reason unable to control their fear. So his aim, using mice, is to find the brain mechanism responsible for inhibiting fear. If he were to find that a certain protein inhibits fearful memories, the next step would be to find a drug for people with PTSD that could encourage the production of this protein. Falls is still working with mice, so this is a long way off, but it's an intriguing approach.

seeking fear

It should be remembered that fear isn't only about immediate danger. In the long-term the fear of losing something we already have, whether a relationship or a job, motivates us to behave in a certain way. If we know we're safe, we can also play with our feelings of fear.

'Our trip into the canyon's going to take about forty-five

minutes. I wanna give you a few safety tips first. If we look as if we're going off the track and about to plunge 300 feet down into the canyon, don't panic. Here in Kiwi-Land we do have an early warning system on board; it sounds remarkably like the opening of the driver's door.' Six backpackers sitting in the back of the truck titter nervously at the joke, longing to relieve the tension, but it doesn't work. They're heading along Skippers Canyon on the South Island of New Zealand. The road is spectacular, cut into the mountain hundreds of feet above the turquoise snake of a river. The trip would be worth making for the views alone, but they barely notice them. That's not why they're here. They're here to experience fear and I'm here to watch them. At the same time, they're trying very hard not to look scared. Some laugh a little over-enthusiastically at the guide's persistently cheerful monologue. After endless hairpin bends they reach a bouncy suspension bridge, built a century ago for gold miners. Next to the bridge there's a set of large, old-fashioned scales. The people queue up to be weighed and lest they decide to lie later on, their weights are written on their hands with marker pens. At the moment they're more concerned about whether it's easier to go head-first or feet-first, or maybe neither.

'Are you all set then?' says Grant the guide. 'Let's throw you off the bridge!' A student from Southampton walks to the middle of the bridge and sits in an old barber's chair under a canopy, accompanied by the blues emerging from a speaker above. Grant wraps a towel around his ankles and attaches a bungee, tugging to see if it's firm. 'Grab hold of your bungee and have a wee look at it. Get ready to say goodbye because it's time to fly, time to check out your headspace. That's it – get close to the edge there.' Now that his feet are tied together the student can only shuffle, but he has to manoeuvre himself under the safety barrier and onto a little wooden ledge. 'Big smile for the camera above your head. Wave to Rich with the video in the love shack over there. A few deep breaths. Don't wait for it to happen. Lots of energy into the dive. Look straight ahead. Take your hands off the rail now and get your wings out like you mean business.'

This individual seemed very jolly in the bus, but now he's gone rather quiet. Of course he doesn't look straight ahead because he can't resist peeping down at the 340ft drop to the turquoise and white river below. Just downstream there's a jet boat surging backwards and forwards waiting to hook him down with a pole once he's dangling upside down just above the water, which will be very soon, assuming he jumps. He swallows and forces himself to look ahead. He

tries to control his breathing. The countdown starts. Everyone else joins in. 'Three-eeee, two-oooo, waaaaahhn, go-ohhhh!' He takes his hands off the rail and bends his knees, but his feet remain on the platform. What he's experiencing is fear. He's aware that he's not really in danger. The bungee has been checked and re-checked, but he is doing something his brain instinctively knows to be dangerous. After the next countdown his instincts are overridden by his determination. He dives out in a huge, slow arc and begins plunging downwards. Far away down below he bounces and soars up again almost to the bridge, then tumbles away from us, then up, then down.

Ten minutes later he has climbed all the way back up the steps to the bridge and he's beaming, desperate to convince me that it was the best thing he's ever done. 'It's just the thrill of doing something that you know you shouldn't. I do something very boring for a living, so when I get the chance I take risks. Every single thing in your body is telling you not to go and then you just go. It's such a rush and then you spot the river coming towards you. I was so scared, but it was brilliant. Now I can't stop smiling.'

That day nearly one hundred people jumped off the bridge. I didn't. I couldn't see why I'd enjoy it and funnily enough when the others are offered a second jump for just a couple of dollars, neither can they. They all insist they will do it again some time, just not today. The guide tells me that 99% of people who attempt it, do jump, but that as soon as one person pulls out, others will too.

This is the safe, clean sort of fear that people deliberately seek out, but it does bring us to the question of what is at the heart of fear. If you know you are safe can you experience genuine fear?

Humans vary enormously in their enjoyment of thrilling activities and in the degree to which they're prepared to take risks. It seems to be a fairly stable aspect of an individual's personality and was presumed to be influenced by a combination of genetic disposition, upbringing and experience, until, that is, a quite bizarre study was conducted in the Czech Republic. *Toxoplasma gondii* is a parasite which can be picked up from undercooked food or cat faeces. It can be dangerous for babies, but the worst that older children and adults suffer is a mild fever and aching muscles. Then it becomes dormant and although the person won't have any symptoms it can stay with them for life. Between 30 and 60% of people worldwide are infected and most of them are unaware of it. It is known, however, that if a rat has this parasite they have slower reaction times and behave more fearlessly than uninfected rats. Jaroslav Flegr, a parasitologist from Charles University in Prague, decided to find out whether the parasite could have a similar effect on humans. He looked back over the records of three years of road accidents and analysed the blood samples taken at the time as part of police or medical investigations. The findings were extraordinary: people with the parasite were more than two-and-a-half times as likely to have been in an accident. Flegr has come up with a possible explanation for these seemingly unconnected facts. Presumably parasites cause rats to become slower and braver because this somehow benefits the parasites, perhaps by increasing the likelihood that the rats are caught by a predator like a cat, enabling the parasite to improve its chances of passing through the food web. Once it infests a human it probably won't get any further because human predators are few and far between these days, but an unfortunate by-product might be that people become both slower and braver.

In modern life the one area where this could make a real difference is the place where we face the most danger – on the roads. It does sound far-fetched and if it is the case that the parasite makes us more likely to take risks then an explanation needs to be found for a mechanism by which the parasite could change our brain chemistry and alter our behaviour. One hypothesis is that the parasite might boost the production of a particular substance in the brain called dopamine. As I mentioned in the chapter on joy, dopamine makes you feel good and is released in large quantities by sex or drugs. At the moment this research is in its infancy. If it were the case that people who have road accidents are releasing more dopamine all the time then they should feel happier than everyone else. Many factors contribute to each road accident and we know that individuals vary considerably in their willingness to take risks, but Jaroslav Flegr believes that if cats were all vaccinated against the parasite then the number of road accidents might be reduced. It certainly is a novel approach to road safety.

the heart of fear

Although we now know something of what happens in the brain and the rest of the body when we feel fear, the question remains as to whether fear is simply a desire to avoid danger coupled with various physiological reactions or is there more to it than that? Do animals feel fear in the same way as we do? When rabbits hop across a field at night, they stand up tall trying to see what's going on, apparently twitching with perpetual anxiety. As soon as they detect a person's presence they flee in terror, but are they feeling afraid in the same way as we do or are they just instinctively running away? Could a frightening experience ever haunt them afterwards? Artificial intelligence expert Dylan Evans argues that it's easy to programme a robot to behave in ways that we'd recognise as fearful. The robot could be programmed to cry out and run away whenever it sees a dog and, watching it, we'd say it must be afraid of dogs. The question is whether that fear would matter to the robot in the same way it does to us.

This is where the issue of consciousness comes in. Are emotions simply a set of useful responses to different situations which we've evolved in order to make quick decisions and survive, or is there something more to them? Surely it is consciousness which allows us to explore an emotion fully?

In an article published in 1884 the American philosopher and psychologist William James declared that emotions were experienced in the body first and that it was an awareness of these bodily sensations which caused us to feel the emotion in our minds. So just as you feel happier if you smile, you watch a film like *Candyman* and as the student looks in the mirror and repeats the magic words your heart starts beating faster. You become aware of your heightened state and then you feel consciously frightened. Or is it the other way around? Maybe you begin to feel afraid and because of that your heart starts racing. Just over forty years after James' work a physiologist named Walter Cannon comprehensively attacked his ideas, stating that physical sensations must be a result of the feelings, not a cause. Do you run away because you're scared or are you scared because you're running?

Recent studies by Hugo Critchley from University College London have found that people vary in their ability to detect changes in the internal state of their body, an ability known as interoceptive awareness. This was only a small study but nevertheless he found that the people who were best at observing and tapping out their own heartbeat tended to be those who generally felt sadder and more anxious. They also had more grey matter in the part of the brain thought to deal with interoceptive awareness. Although our interoceptive abilities are not something we tend to be aware of – we can't know what other people's heartbeats feel like – previous research has suggested that the more body fat people have, the more difficult they find it to detect their own heartbeat. If the ability to feel your heart beating does affect anxiety, this prompts the intriguing idea that fatter people might be protected against feeling anxious or scared.

The debate over whether emotions precede physical feelings or vice versa continues. No one can be sure of the answer, but maybe it happens in either direction, depending on the situation. It does seem to be the case that our bodies can have more of an impact on our emotions than we tend to think. It has been assumed that our consciousness and therefore our awareness of the experience of fear is located solely in the brain, but after considering the experiences of people with locked-in syndrome the neurologist Antonio Damasio believes this might not be so. This is a condition where a particular part of the brain has been damaged, blocking messages from the brain to the muscles, leading to paralysis of the entire body apart from vertical movement in one or both the eyes. People with this syndrome are fully conscious and aware of their situation, but usually their only means of communication is through moving their eyes. Jean-Dominique Bauby dictated the book *The Diving Bell and the Butterfly* by blinking his left eyelid while the alphabet was repeatedly read out to him. People often marvel at how well these patients come to terms with their situation, but Antonio Damasio believes this may be due to the nature of consciousness. He has noticed that they don't usually experience the fear and anguish that you might expect at being unable to move, while patients with other conditions which force them to remain still, even temporarily, become far more distressed and frightened. His theory is that although they experience emotions, these feelings don't have the same impact because they are not experienced in the body, which could explain why people with this syndrome apparently cope so calmly. This suggests that bodily feelings do play a part in the way we experience fear and that the body and brain send messages to each other, reinforcing what we feel in some kind of feedback loop.

Neuroscientists have made more progress with fear than with any other emotion, but so far their research has demonstrated the complexity of the system for this supposedly basic emotion. During fear the brain focuses in on details which might ensure our survival whilst taking account of context, resulting in an emotion which can be life-saving or leave us literally scared to death. Although the subject of our fears changes as we grow up, fear appears to be more on a knife-edge than other emotions; too little can put us in mortal danger, while too much can lead to long-term trauma.

There are occasions where we deliberately seek out this apparently negative emotion. Control seems to be the key which can make the difference between an event which is thrilling or traumatic. Fear and memory appear to be linked in extraordinary ways and in years to come we can expect more discoveries about the ways in which the production of proteins and the activity of neurotransmitters impact upon our experiences of fear, our memories of those fears and most importantly on what is happening when the system goes wrong. Who knows, maybe if more is learnt about the inhibition of our fear one day we might be able to have more control over this feeling, choosing when to feel scared and, just as importantly, when not to.

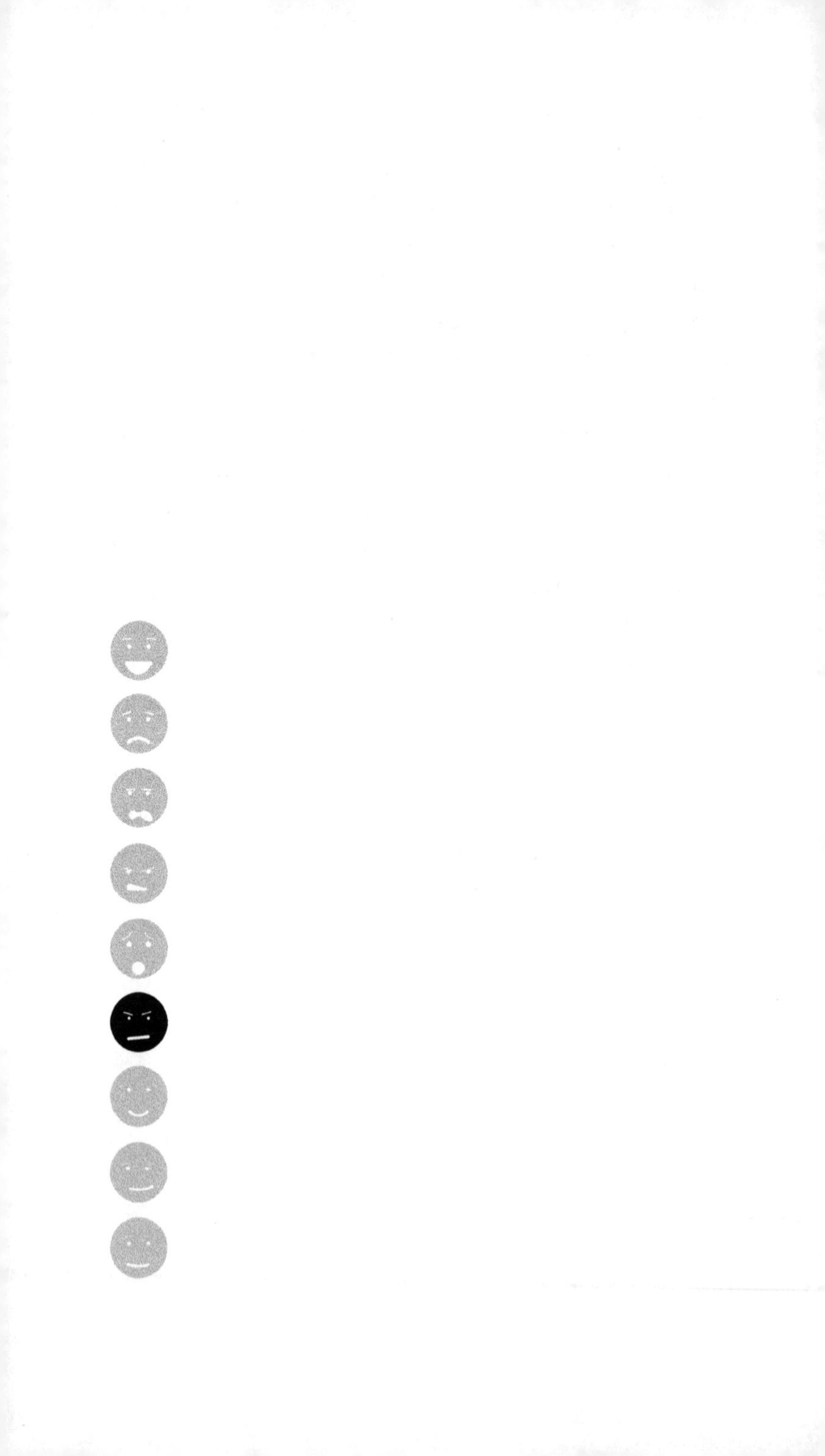

six

jealousy

In a Spanish village farming peasants are celebrating a wedding. With a flower garland in her hair, the bride and local beauty, Casilda, dances barefoot with her groom, Peribanez. They are clearly in love. Later they happen to meet an injured Commander who is passing through the village and the kindly Casilda tends to his wounds. He soon becomes obsessed with her and plots his return to the village to seek her out while Peribanez is away. She's not interested, but when Peribanez hears of the Commander's obsession with his wife he is overcome with jealousy, storms home and finding him there stabs him to death.

Matilda and her husband Philemon sit eating dinner in their home in a South African township. On the third chair there's a suit, laid out carefully across the front of the chair, as though there were a person inside. Nervously Matilda pretends to feed dinner to the suit while Philemon watches. This is her punishment. Philemon was renowned for his kindness until he returned home unexpectedly one morning and discovered that a man was making daily visits to the house while he was out working. When the man was discovered in bed with Philemon's wife he fled naked, leaving behind his suit. Now the suit must accompany them wherever they go. By threatening to kill her, he even forces her to carry the suit along the main street when they go out for a Sunday stroll. He can never let her forget what she's done. She joins a church group and finds new friends, determined to make her marriage work and to fill the long days while he works. She feels much happier and months later she throws a party to thank her new friends, but Philemon still can't forgive her. He is not going to let her forget her past wrongs. Waiting until the party is in full swing he forces Matilda to introduce the suit to each and every one of her new friends. Then he abandons the party and goes out drinking, unaware that on his return his life will change forever. When he eventually gets home from the bar he finds Matilda lying on the bed. She is dead.

These are the plots of two plays which happened to be on at the same theatre within the space of a few months. Four centuries separate the lives of the characters in the two plays but the same emotion transforms two apparently reasonable men and in both cases the result is death. The emotion is jealousy. *Peribanez* by Lope de Vega is thought to have been written some time between 1605 and 1614. *Le Costume* written by Can Themba is set in a South African township in the twentieth century. Everything surrounding the characters' stories is different, but the jealousy is the same.

As an emotion jealousy is unique because not only does

it feel unpleasant, but we hate ourselves for feeling it at all. We all know it's an ugly feeling which eats you up from the inside. As Shakespeare put it, jealousy 'doth mock the meat it feeds on'. The French for jealousy – 'jalousie' – is the same word used to refer to Venetian blinds, perhaps reflecting the sense that nobody wants to be seen experiencing jealousy or that the jealous individual spies through the slats of a blind to check on their partner's behaviour. I imagine a man in the street waiting outside his girlfriend's flat because he is suspicious. He sees another man go into the flat. It's night and the light is on so he can see them moving around. First one figure flicks across the window, then the other. What are they doing exactly? He never sees them both at once. Then she comes over to the window, briefly looks down at the wet street and with a quick twist the Venetian blinds are closed. Darkness takes its place in the window. Now he can only stand outside, torturing himself with his imagination of the activity taking place behind the blind.

Writing in the sixteenth century the essayist Michel de Montaigne said of all the emotions, 'Jealousy and envy her sister seem to me to be the most absurd of the bunch.' He was particularly concerned about the effect jealousy had on women (he wasn't known for his complimentary views of women, disliking their moaning to such an extent that he applauded a man who said that for a marriage to work, you needed a blind wife and a deaf husband). He wrote, 'When jealousy seizes hold of the feeble, defenceless souls of such women it is pitiful to see how it bowls them over and cruelly tyrannizes them. It slips into them, under the title of loving affection: but as soon as it gets possession of them, those same causes which served as a basis for benevolence now serve as a basis for deadly hatred. Of all the spiritual illnesses, jealousy is the one which has more things which feed it and fewer things which cure it.'

Jealousy is a complex emotion comprising elements of fear, anger, sadness, anxiety and despair. In addition it brings the shame of suspecting your loved one and turning them into an object of hate. While people admit to feeling unhappy or worried about their relationship, they tend to avoid any mention of jealousy. We know it's unattractive.

When people experience extreme jealousy they report feeling hot, nervous, shaky and empty in the stomach. Despite the existence of these physical markers, there is no facial expression for jealousy. Darwin said this is why painters 'use accessories to tell the tale' of jealousy while poets resort to 'vague and fanciful expressions' to describe the feeling. Green is often used in descriptions of jealousy, but it's not clear why. The words 'green' and 'pale' share the same word in classical Greek which might have led to a mistranslation into English. However, jealousy was once associated with an overproduction of bile, giving the face a greenish hue.

Alex was eighteen when a colleague at the restaurant-club complex where he worked introduced him to an older woman. Feeling she had missed out by marrying a man in the navy when she was very young, the woman decided to make up for it while her husband was away at sea. Alex was thrilled to have an older woman pay him so much attention and soon they began an affair. Each night she would sweep up to the club in her flashy car and take him home with her. After her husband's ship returned to port the affair continued and one day Alex even found himself cycling round to the house when he knew she was out and ringing the doorbell. Recognising him as a friend of his wife's from the club, the husband invited him in and confided to Alex that his marriage wasn't going well and that he suspected his wife was seeing someone else. Knowing very few people in the area, he said he had no one with whom to discuss it. Alex comforted him as best he could and then left, filled with guilt.

Some weeks passed and the woman told Alex she had bad news. She was giving her marriage another try and she and Alex must stop seeing each other; her husband was in some sort of trouble at work and might even be court-martialled. The next time Alex saw her was two years later when he happened across her in the street. This time she was more forthcoming about her husband's difficulties at work. Discovering Alex, his confidant, was the man sleeping with his wife, he had made a plan. He took an assault rifle, loaded it and abandoning his post guarding the armoury he waited in the bushes outside the club, listening for the sound of his wife's car as she came to pick up her young lover. How dare he sit there pretending to be sympathetic while he poured out his heart to him about his problems with his wife? How they must have laughed together about him afterwards, but they wouldn't laugh together anymore. He would kill them both. The jealousy he felt was so strong that he was prepared to kill the woman he loved, destroy his naval career and inevitably spend years in prison. The car arrived. His wife disappeared inside the club and moments later she emerged laughing with Alex – always laughing. Now was his chance. Alex would be first. He got him in his sights and was ready to pull the trigger. But he could not do it. He was not going to let jealousy overcome him.

Sexual jealousy certainly can drive people to kill; it's implicated in up to 20% of cases where men murder other men, although in Western cultures the women are more often the victims of the violence than the man's rival. This is jealousy at the extremes. Nevertheless milder jealousy can still transform a person's behaviour. They might follow their partner, monitor their phone calls and check their pockets, desperate to catch them out, whilst simultaneously dreading the successful discovery of infidelity. One man I know was so jealous when his girlfriend went to live abroad that he took a magnifying glass with him when he visited and while she was out at work he scanned the walls for drawing pin marks – evidence that she might have removed a photo of her lover prior to his arrival. A journal article even describes a woman who marked her husband's penis with a pen and then checked it each evening to see whether it had been touched.

Jealousy this extreme is rare, but temporary jealousy – perhaps while watching a partner dancing, flirting or deep in conversation with someone else – is far more common. Despite this, a man called Boris Sokoloff reports that back in 1948 he was unable to find a single article in either law or medical libraries which dealt with the subject. In an attempt to fill that gap he wrote a surprisingly entertaining book on the subject, containing a succession of tales of jealousy. For example, there's the 'Case of the Doctor's Wife' which tells the story of Mrs P whose first husband commits suicide after she leaves him to marry a doctor. All too aware of the bewitching effect of the doctor, Mrs P begins to worry that other women might fall under his spell. 'I was jealous of him from the first day of our marriage, perhaps long before. And how could I not be jealous? He was fickle and changeable, and so attractive. And I thought that if even I, a woman of considerable force of character, fell so easily a victim of his charm, how could other women resist him?' Eventually she becomes so convinced that her husband might be having an affair with his attractive secretary that on overhearing the secretary telling him that she is pregnant, Mrs P assumes that her husband must be the father. She takes his gun, waits at his office and shoots the secretary dead in an attempt to frame him for murder.

She didn't get away with it, but as recently as the end of the nineteenth century jealousy was considered a legitimate defence for murder. It was an emotion that was even admired. Only later did jealousy begin to be seen as a sign of insecurity that should be kept hidden.

By the 1960s people were experimenting with the possibility of eliminating jealousy from relationships altogether. There was one group who not only accepted the knowledge of their partners' infidelity, but would happily witness it, whilst apparently feeling no jealousy – swingers. From 1969 Brian Gilmartin spent three years comparing swinging couples with non-swingers. He found that swingers avoided jealousy by separating their recreational sex with a number of people from the romantic sex they had together. Sex with other people simply became a shared hobby. However, jealousy was not always successfully eliminated. He found that when a new couple came to a swinging party, it tended to have been the man's idea. The men were often so excited about the idea of sleeping with other women that they forgot that they would also see their own partner with someone else. Jealousy would often take them by surprise, Gilmartin said. 'If, as often happens, she enjoys the experience, his expectations are upended and he loses his balance. Jealousy attacks him in his guts with the force of a strong blow.' Meanwhile experienced swingers reported feeling only mildly jealous and, rather than reacting with rage like some of the newer swingers, the jealousy would add to the sexual passion they had for their partner when they returned home after the party. Interestingly he found swingers to be happier in their marriages than the non-swingers he examined, but if jealousy was to be avoided a secure relationship was needed.

Jealousy is not a simple emotion to study. The observation of jealousy in public places isn't easy, nor is induction of the feeling in a laboratory, but a researcher called Krystyna Stryzewski-Aune made a good attempt in 1993 when she invited dating couples to take part in an experiment. After each couple arrived in the laboratory they were told that the video camera wasn't working and were left to wait with an attractive assistant, sometimes male, sometimes female, while it was mended. With half the couples the assistant would chat pleasantly, but with the other half the assistant would flirt openly with the partner of the opposite sex, flattering them, leaning close, accidentally brushing a hand against their leg and finally asking for their phone number so that they could call them with the results. As soon as the flirting began, both the women and to a lesser extent the men would clearly signal ownership of their partner by, for example, rubbing their backs, patting their chests or brushing something off their shoulders. The jealous women kept smiling and talking pleasantly, while the men were more likely to withdraw from the conversation and scowl.

More recently Paul Stenner from University College London interviewed couples to see how they make sense of jealousy. Intriguingly, he found a recurrent theme was for one person in a couple to characterise their partner as a jealous type, saying they were careful not to inflame that jealousy. Meanwhile the other person would claim not to be jealousy-prone, nor in need of molly-coddling. Despite these apparently contradictory viewpoints Stenner found that ideas about jealousy seemed to play a useful part in their dealings with each other. Some couples in his study told him that jealousy added spice to their relationships and that the awareness that other people were attracted to their partner encouraged them to put more effort into making the relationship good. In extensive studies on jealousy Ayala Pines in California has found that an individual's opinion of jealousy depends on whether it's an emotion they tend to feel themselves. People prone to jealousy see it as a positive emotion and tend to like their partners to be jealous, while non-jealous people see it as a sign of immaturity. Surprisingly, the jealous people saw the emotion as part of their own personality, rather than blaming it on their partner's behaviour. She also found that people who believe in monogamous relationships are less likely to be jealous while those who have had affairs themselves are more jealous.

Unlike some emotions, jealousy can be employed strategically. If you tell your partner that someone asked you out it's a gentle reminder that there are other people out there who are interested in you. This is the subtle approach. Sometimes people will choose more obvious tactics and spend parties deliberately flirting with other people which risks their partner attempting to monitor their future behaviour. Jealousy can even ignite the partner's sexual interest, enabling them to see their partner through the eyes of a stranger. The psychologist David Buss from the University of Texas, who has done a great deal of research on the subject of jealousy, tells the story of how he used to play tennis regularly with an attractive married woman who told him that she always had great sex on tennis days because her husband was slightly jealous of him.

If mild jealousy can be beneficial for a relationship, then you might expect jealous couples to stay together for longer. One study set out to discover whether this was true by assessing couples' jealousy and then following them eight years later to see how they were getting on. In fact the most jealous couples were the ones still together. This has been taken as evidence that jealousy holds relationships together, but there is the possibility that the couples who were the least jealous simply weren't that bothered about whether the relationship continued, in which case it's not surprising that they would be the ones to split up.

To an extent there is a correlation between jealousy and love; if you don't love somebody at all you might not feel threatened by the idea of them going off with someone else. The same does not apply at the other end of the scale. A highly jealous person is not necessarily any more in love than someone who is less jealous. After studying different societies around the world, the anthropologist Margaret Mead concluded that jealousy was experienced to the same degree worldwide. She said that jealousy 'is not a barometer by which depth of love can be read, it merely records the degree of the lover's insecurity'. Her theories have been borne out by several research studies indicating that people with lower self-esteem experience more jealousy. This is plausible. If you think that you are not worthy of your partner's attentions and that they would find most people preferable to you, then you are likely to feel jealous. Back in his 1948 research on the subject, Sokoloff was convinced it was not love, but fear of isolation that was responsible for irrational jealousy. 'A jealous person cannot face the world alone. He must have someone to share his loneliness with him, his partial or complete isolation from the outside world.' More recently research has suggested that as well as feelings of inadequacy leading to jealousy it could happen the other way around – knowledge of your own jealousy can make you feel inadequate. This fits in with Karl's experience: 'At various times in my life with different people I've experienced jealousy, sometimes in an irrational way and at other times with fair cause, to the point where I've become morose, depressed with very, very low self-esteem – feeling unlovable and physically unattractive. It's even made me incapable of holding down a job at certain times. If one person who has stated that they're in love with you can betray you in that way, then so could everybody else. You feel as though you are not worthy of love. There are those dark nights of the soul where you're sitting there imagining that she's with someone else – it's almost like a madness. Unfortunately people are more likely to be unfaithful to you when you are at a low point and feeling unattractive, so it's not a time when you're likely to say I can deal with this. It causes great sadness and grief and enormous anger. The only time I've ever been violent in a relationship with a woman was in relation to sexual jealousy.'

In an attempt to explain the relationship between self-esteem and jealousy, David DeSteno and Peter Salovey from Yale University have looked at the ways in which we constantly compare ourselves to others. We spot the areas in which we are lacking and others excel. If your friend is successful in an area of life in which you are not involved, you can bask in their glory, so being good friends with a famous actor or sportsperson, for example, is fun if you are in no way in competition with them. The nineteenth-century psychologist William James said that he would be perfectly happy to meet the world authority on ancient Greece, but wouldn't want to meet someone who was a better psychologist than he because that might threaten his own evaluation of himself. Following this line of thinking, DeSteno and Salovey guessed that people would feel more jealous if their partner were to flirt with someone who had the attributes they most valued and would most like to have.

They tested this by finding out which attributes individuals found the most important and then asking them to imagine a series of hypothetical situations in which their partner was flirting with someone who was either very popular, very intelligent or very athletic. As predicted people felt most jealous when their partner was with someone with the very talents they most admired. There was also an interesting difference between men and women. The men tended to be the most jealous of the rival who had the characteristics that he most admired himself, while the women were the most jealous of the women who had the characteristics they thought their partner desired in a woman. What's intriguing is the way that this all changed when women were asked to rate the other women not as potential rivals in love, but as possible friends. Then the women of whom they were the most jealous became the same women they would most like to be their friends.

When Ayala Pines asked people how jealous they would feel if they discovered their partner was having an affair, people felt more jealous if they knew the lover, but surprisingly it made no difference whether that person was their friend or merely an acquaintance.

So far, jealousy seems not to be based on the strength of our love, but on our feelings about ourselves. However, there is another, more bizarre explanation as to why some people suffer from jealousy more than others; it might involve the length of their ears.

wonky features

Fifty students lined up in a corridor at Dalhousie University in Nova Scotia, waiting to take part in a study. One at a time each student went into a room and removed their shoes. Then a researcher took digital callipers, fitted them carefully around the foot and noted down the width measurement.

The same happened with the fingers, ankles, hands, wrists and even the elbows. Lastly the researcher, who had no idea how these dimensions were to be used, noted down the length of each ear. Then a second researcher came in and took all the same measurements again. Afterwards each student's tendency to feel jealous was assessed using a standard questionnaire scale plus two supplementary questions: how jealous would you feel if another person got the promotion for which you were qualified or if another person was praised for something for which you were in fact responsible? The students must have wondered how on earth the width of their elbow could possibly bear any relationship to their tendency to feel jealous when their partner chats up someone else at a party, but apparently it does.

Extraordinarily, the most jealous people were found to have the least symmetrical bodies. If you have one ear a lot longer than the other the chances are you are someone who gets very jealous. There will of course be exceptions, but across a large number of people, the least symmetrical are the most jealous. At first sight this seems ridiculous, but it's not that your uneven ankles are somehow communicating with your mind, impelling you to start searching through your partner's pockets for clues to infidelity. William Brown's research is based on the idea that symmetrical features tend to be considered the most attractive. Clearly we don't go round measuring people's limbs to judge their attractiveness, but if we did we'd often find that the people we rate as the best-looking are also the most symmetrical. This makes evolutionary sense because asymmetries are caused by genetic mutations in development. Therefore if you are looking for a person with whom to mate, a more symmetrical person might be more resistant to such genetic randomness and therefore produce stronger children. Thus this study is really concerned with attractiveness rather than symmetry *per se*. Oscar Wilde noted the inverse relationship between jealousy and good looks, when he said, 'Plain women are always jealous of their husbands. Beautiful women never are.'

The theory, then, is that the less you have to offer your partner, the more likely you are to worry that they will choose someone else over you. When it came to jealousy of colleagues, the asymmetrical people felt no more strongly than anyone else, suggesting this phenomenon is specific to sexual jealousy. Feelings of jealousy enable the person to monitor and if necessary improve their relationship to encourage their partner to stay. In theory the most symmetrical people don't need to do this because their partners are supposedly more likely to stay due to the good symmetrical genes already available to them.

The theory is all very well, but in real life when are situations ever this simple? Presumably, if symmetry is this important then both partners' measurements need to be taken into account. If you are asymmetrical, but so is your partner, then you might not need to worry that they will find someone more symmetrical because the other person won't choose them. The other problem is of course that symmetry is not the only factor in attractiveness nor is a person's physical attractiveness the only attribute a person considers when they decide whether to stay or leave. People might well leave a relationship because they find someone better, but they are not necessarily better-looking; they could be kinder or funnier or more intelligent.

When it comes to attractiveness it's the degree of match between couples that matters. David Buss has found that when there's a mismatch jealousy is more likely to occur. If you are both a '6', say, that's OK because you both know the other person probably couldn't do a lot better than you; they would be foolish to leave and risk finding no one superior. However if you are a '6' with the good fortune to have fallen in with a '9' you might be slightly more concerned that they will find someone better and thus more prone to jealousy.

Looks can't explain all jealousy. Circumstances inevitably play a part, particularly across the lifespan. The cliché is the jealous full-time mother exhausted by looking after her toddlers, aware that she is ageing, while her increasingly successful husband is surrounded by women at work who don't have disrupted sleep and are free to stay out late. Later on in life these roles can swap as many men become infirm before their partners and while they become housebound, jealousy of their wife can increase because she is free to 'gad about meeting who knows whom'.

when jealousy becomes a problem

Once they discover a rival, people respond in various ways. They might refuse to acknowledge it, control their partner's movements or improve their own physical appearance in order to compete with the rival. Once people feel jealous they often start behaving in less constructive and less appealing ways, driving their partner away from them. Using role-plays one researcher found that when people were feeling jealous they behaved coercively towards their partners, trying to force them to bend to their point of view. When they weren't feeling jealous they were better at finding resolutions with their partner.

It was certainly Karl's experience that jealousy rendered him unable to communicate with his partner. 'When people talk about the green demon they're talking about possession and I do believe jealousy is exactly like that. All of my normal mental processes are completely unoperational at that point. There's no sense of rationality, no sense of proportion. Quite often the imaginings are far more horrific than the actuality, but one of the problems for me in finding out my partner has been unfaithful is that I've felt immensely stupid. The whole world knows except for me. I remember taking my boys to school and being taken to one side in the playground by well-meaning women to be told about my partner having an affair with someone and me laughing at them, utterly convinced that my partner would do no such thing. When I found out it was true my feeling of stupidity and deep embarrassment was profound. I had to continue to take the kids to school which just compounded it because the particular guy involved also took his kid to the same school each morning so we had to meet up and it was witnessed by all the other mothers.'

In this case Karl didn't suspect his partner, but Simon Gelsthorpe, Consultant Clinical Psychologist at the Bradford District Care Trust, has found that when people do want to question their partner's behaviour, jealousy often drives them to express themselves angrily, accusing their partner of things that might not be true. An alternative approach, he suggests, is to express jealousy as a fear. So instead of accusing your boyfriend of flirting with other women you say, 'When I see you talking to beautiful women all evening I feel afraid that perhaps you would like to be with them instead of me.' The idea is that this opens up a dialogue, rather than inciting a huge row. Some people will deliberately befriend the rival or emphasise the rival's weak points. Others will play on a partner's guilt to force them to stay or at the extremes they turn to violence.

Those researching jealousy are faced with the problem that they cannot know whether a person has good grounds for feeling that way or whether the emotion is irrational. With other emotions such as fear, the person themselves usually knows that a phobia, although hard to control, is irrational, but with jealousy it is more complex. Some use the term 'morbid jealousy' or in psychiatric circles 'Othello syndrome' to describe jealousy which is unfounded. However, defining when jealousy becomes excessive can be a subjective judgement which varies between cultures.

morbid jealousy

Morbid jealousy can sometimes occur as part of another condition such as schizophrenia or organic brain disease. There are case histories of jealous delusions in people who have suffered a stroke in the right hemisphere of the brain. In 1999 a case of Othello syndrome was reported in a woman who had made a good recovery from a stroke at the age of twenty-five, but five years later became obsessively jealous and was so depressed that she took an overdose. After treatment with anti-depressants the jealousy subsided, but her doctors believed that brain damage from the stroke was the cause.

There also appears to be a physiological link between alcoholism and obsessive jealousy. This was reported back in 1905, but the relationship is by no means straightforward. Many people with alcoholism show no jealous impulses. Moreover it's hard to disentangle jealous outbursts from the disinhibitory effects of alcohol which might lead people to voice feelings that others keep to themselves. Perhaps the drink confuses their thoughts and leads them to draw false conclusions. Alternatively it could happen the other way. Perhaps fear about their partner's fidelity is one of the reasons they turn to drink. There is also the possibility of course that they are correct and that their drinking has driven their partner to look elsewhere.

Whether or not jealousy has any foundation in reality, it changes the way people think. A morbidly jealous person begins to base their views on erroneous beliefs, for example that everyone finds their partner attractive and is actively pursuing them. Every observation is misinterpreted as 'evidence' against their partner – the car which drives away quickly just as they arrive home or their partner's tardiness in coming home from work. As Karl experienced when he had good reason to feel jealous, thoughts were constantly interrupted by images of what their partner and new lover might have been doing.

People with morbid jealousy often describe feeling depressed and anxious, even suicidal. In a study of people in the UK suffering from morbid jealousy, Mairead Dolan and Nagy Bishay found that most felt that jealousy was ruining their lives, despite the fact that more than half were aware the feelings were unfounded. Half had spied on their partners or attempted to entrap them, while more than a third had considered hiring a private detective.

Albert Ellis, the creator of Rational Emotive Therapy, believes that we can choose whether or not to feel jealous. Even if your partner leaves you for someone else, they have not inflicted jealousy upon you. He strenuously insists that it is up to us to decide how to react. '*Because* my mate is carrying on a hot affair with So-and-so, *that* makes me jealous and hurt. Bullshit! There is no magic by which any outside event, such as your spouse's being intensely involved with another person, can wriggle its way deep into your gut and hurt or upset you.'

Once again it is our belief system that dictates the way we feel. The theory is that if you believe that it's dreadful that your partner is seeing someone else and must be a bad person for doing so, then that is a judgement you are making. If you decide that these things are true you will feel jealous, but if you decide to look at the situation differently then you won't. This is, of course, easier said than done. To be fair, though, Ellis did put his own theories into practice during the late 1930s when he was living with a woman whom he knew to be cheating on him. He would stay awake until the early hours waiting for her to come home. Then, looking to the writings of philosophers for comfort, he decided that if this was how she was, then that wasn't his fault and he would accept her as she was. He even greeted her happily on her return, and asked her to recount her exploits before going to bed with her himself.

does jealousy have a purpose?

Of all the emotions, jealousy is the one which most seems to lend itself to explanation within evolutionary psychology. The theory is that every man's priority is to pass on his genes. Therefore he needs to ensure that his partner doesn't spend valuable time pregnant with someone else's child, nor that he wastes resources on looking after a child that he can't guarantee is his. In order to protect his genetic heritage feelings of jealousy arise if he sees his mate interacting with another man in a way that might lead to sex. The idea is that sexual jealousy in men is hard-wired. So even if they know their partner is on the pill and can't get pregnant by another man with the resulting cost to their resources, they will still feel jealous if she flirts with someone else.

These theories tend to suggest that monogamy is important only to men. The idea is that if a man sleeps with another woman it doesn't have serious consequences for his original partner because he would still protect and provide for his own children. In contrast a deep emotional relationship with another woman could constitute a real threat to his partner's children because he might abandon them in order to begin a full-time relationship with the new woman. This explanation is used to account for the fact that men supposedly become more jealous at the idea of sexual infidelity in their partners, while women are more worried that their partner might fall in love with someone else. Of course today women can bring up children alone but within evolutionary theory the argument goes that this is a recent change, not manifest in our emotions which have taken thousands of years to evolve. In theory, random changes in the way that people think, even, continue to be passed on through the centuries if they happen to be useful due to survival of the fittest. Thus the theory is that the men who experienced sexual jealousy and the women who experienced jealousy of their partner's emotional relationships were the people who successfully produced offspring who made it to adulthood. However, there are in fact distinct advantages for women in ensuring that their partners do not even have purely sexual relationships with other women because if they become pregnant both resources and attention would be diverted away from her children.

One of the major proponents of the idea that men and women experience jealousy differently is David Buss from the University of Texas. Buss reached his conclusions through a method he calls 'Sophie's Choice', named after the film where a mother was forced to choose which one of her two children would be saved from certain death. In fact Buss's choice is rather less dramatic; people are asked to imagine a committed sexual relationship which is either past, present or future. They have to decide whether they would be more upset if their partner had sex or developed a strong emotional bond with someone of the opposite sex. When given this choice more women rate the emotional infidelity as worse, while more men are upset about the sexual infidelity. However, although more men chose sexual infidelity it was by no means all of them. If this were an innate sex difference which has evolved, you would expect fewer exceptions than there were.

This work certainly provides a neat explanation for the ways in which men and women sometimes behave differently in their attitudes towards relationships. However, the research is not without its problems. Firstly, people are required to make a choice about which type of infidelity is worse. There is no opportunity to say that both would be upsetting or that it depends on the circumstances. This forced choice somehow pushes men towards one decision and women to the other, but when David DeSteno from Northeastern University in the United States asked people to rate the two options separately using seven-point scales (ranging, for example, from not jealous through to incredibly jealous) the results were rather different. Regardless of the order of the questions both men and women rated sexual infidelity as more upsetting. One explanation for these strikingly different results is that choosing between two forms of infidelity leads to a more complex decision-making process, forcing people to weigh up the pros and cons of both at the same time. To test this DeSteno went back to the Sophie's Choice method, but just before people answered the infidelity question they were given a string of seven numbers to memorise for later. This well-established method of distraction prevents the brain from processing information at as deep a level as when a person concentrates on just one subject at a time. Suddenly the sex difference disappeared, with women as much as men identifying sexual infidelity as the most upsetting.

This is trumpeted as disproving Buss's theory, but there

is one problem. The situation is an artificial creation. What are the chances of discovering that your partner has had a one night stand and that while you're considering their explanation that they don't love the person you happen to be distracted by another task such as memorising a list of words? If women tend to process this hypothetical problem at a complex level, then they would be expected to do so in real life as well. Moreover, why should women use complex reasoning, when men apparently do not? Having said this, Sophie's Choice also lacks reality. It's unlikely that our ancestors were faced with both sorts of infidelity simultaneously, forcing them to make a judgement about which was worse.

In one of the most severe and comprehensive critiques I've read of a person's work, Christine Harris from the University of California, San Diego, takes Buss to task for, amongst other things, the use of hypothetical questions. Although you can ask people to guess how upset they would be in a certain situation, until it happens, no one can be sure how they would react. The people in these studies are usually undergraduate students, some of whom won't yet have experienced these sorts of dilemmas. When Harris looked at genuine situations people's reactions were often very different.

However, Buss's theories don't only rest on the answers to hypothetical questions. He also provides physiological evidence. In the laboratory he measured heart rates, perspiration levels and frequency of frowning in men and women while they imagined either that their partner had a one night stand or was becoming emotionally involved with someone else. Men's heart rates were almost five beats a minute faster when they imagined their partner's sexual infidelity compared with emotional infidelity. With women you would have expected it to be the other way around, but in fact their heart rates were the same for both types of infidelity. The predictions were borne out, however, when it came to sweating. Men showed a larger sweating response to sexual infidelity, while women sweated more at the thought of emotional infidelity. Unfortunately this is hard to interpret because jealousy is not the only emotion to provoke a person to sweat, nor to make their heart go faster. Fear has the same effect, disgust increases sweating while both anger and happiness speed up the heart rate, as does another feeling, which might be particularly pertinent to Buss's work – sexual arousal. If a man's heart rate rises when he imagines his partner with another man perhaps he is feeling not jealous, but turned on. Christine Harris set out to discover whether this was the case by giving men two scenarios to imagine – one sexual and one emotional. Half pictured themselves in the scenes with their partners and the other half imagined another man in their place. She found that the physiological reactions noted by Buss could occur simply through imagining a sexual scene. Men's heart rates and blood pressure rose in both the sexual scenarios regardless of their opinions of sexual and emotional infidelity.

So after all this research it seems that women do get more upset about emotional rather than sexual infidelity, but only in two very specific circumstances – when they can focus their whole attention on the question and when they have to make a choice about which of the two types is worse rather than rating the severity of each separately.

Not every culture responds to jealousy in the same way. In some cultures at certain times in history it has been perfectly acceptable to sleep with your husband's brothers. At the other extreme are societies where adultery is viewed as so appalling that the adulterous spouse is killed. Evolution has clearly not given us a universal reaction to infidelity. Gary Brase and his colleagues at the University of Sunderland gave the Sophie's Choice question to people from England and Romania and found that only a third of Romanian men chose sexual infidelity to be more upsetting, compared with 52% of the British men.

It tends to be assumed that, even if life today has changed, in earlier societies it was useful for men to experience sexual jealousy and for women to experience emotional jealousy. It is true that in order to pass on their genes men needed to know that their partner's children were definitely theirs. However, it could be argued that a good way to ensure their partner only has sex with them is to watch for the development of any close emotional relationships with other men. These could be prevented before they led anywhere. By the time a man catches his partner *in flagrante delicto* with another man, it could be too late; they could already be pregnant.

Murder statistics are sometimes cited in support of innate differences between men and women in the experience of jealousy. More men than women murder their partners out of sexual jealousy. However, there are problems with these assumptions. This does not indicate that men mind more about their partner's sexual infidelities. It is simply the case that most murders within relationships are carried out by men. Fewer women rob banks than men, but that doesn't prove that they are any less interested in money. After analysing every study she could find on the subject, Harris concludes that men are simply more likely to respond to a situation with violence and that male violence is more likely to result in death, whether or not it was born out of jealousy.

The way to judge the murder figures more accurately, is to take all the cases where a man murders his partner and all the cases where a woman murders her partner and then see which has the higher proportion of murders inspired by jealousy. When Harris examined twenty studies looking at murders in populations from Africa to Scotland there were almost 4,500 murders committed by men and 800 by women. She found that although there were far fewer murders by women overall, jealousy was at the root of the same proportion of murders by women as by men.

Moreover if it were the case that men were hard-wired with an instinct to kill unfaithful partners, surely we would see far more murders. In fact cases that end in murder are so rare and extreme that they might have nothing to reveal about everyday jealousy.

jealous babies

Sexual jealousy between adults may be painful and at times destructive, but it is by no means the first time people experience jealousy. It starts long before any thoughts about sex, maybe even before we can walk.

A century ago it was assumed that in the same way that

adults and children feel jealous, so could babies. Then, led by Jean Piaget, developmental psychologists began to understand how children's thinking develops and that they don't think as mini-adults. Since then it has been assumed by most researchers in the field that jealousy is too complex an emotion for babies to experience. To feel jealous that your mother is paying more attention to another baby you need to understand that the other baby is a different person with their own experiences – known as a sense of self. Nowadays it tends to be assumed that jealousy doesn't emerge until the second year of life, but some psychologists including Riccardo Draghi-Lorenz from Surrey University, who happens to be the grandson of the famous ethologist, Konrad Lorenz, believes that babies understand more about their relationships than we realise. As part of some research on shyness, he was watching a video of babies when he noticed that sometimes the babies appeared not to be shy, but jealous. To explore the idea further he sent questionnaires out to parents and some were indeed able to describe instances of jealousy in their babies. One mother even invited him to come and witness the jealousy for himself. In a home experiment she paid attention to a friend's baby while ignoring her own. Although theoretically too young to experience jealousy, the nine-month-old attacked the other baby.

Draghi-Lorenz then set up a more systematic experiment. For five minutes a mother would talk to Draghi-Lorenz while ignoring her infant. Then for the next five minutes she would kiss and talk to another baby. Out of twenty-four babies just three cried while the mother talked to Draghi-Lorenz, but thirteen cried as soon as she made a fuss of another baby. The problem is, of course, that by the time the babies were in the second situation they had already been ignored for five minutes so perhaps it was the fact that they were ignored for another five minutes which upset them rather than the attention the other baby was receiving. To overcome this, his latest research was slightly different. This time the baby sat opposite their mother and another woman. In one situation their mother held the stranger's baby and in the other the stranger held her own baby. The infants seemed to enjoy watching the other woman play with the baby but became upset if it was their own mother showing love and affection for another baby. Of course, it is hard to know whether this was typical of these babies' behaviour or whether they were just having a bad day or feeling unwell, so Draghi-Lorenz has also followed a group of babies from birth, visiting the parents every couple of weeks to see which emotions they had witnessed. One mother noticed her baby was jealous at just two-and-a-half months. Draghi-Lorenz believes that babies are capable of feeling jealousy at any age, but that in the first few weeks when their eyes are unable to focus on distant objects, they are unlikely to see anything which provokes their jealousy. Moreover, until they have developed a strong bond with their parents babies might not mind them lavishing attention on others. In fact babies with depressed mothers do not display jealousy, because they do not have such a strong bond with their mothers in the first place. Meanwhile jealousy is more noticeable once a baby has the freedom to crawl across the room and hit the object of their jealousy.

There is also the possibility that jealousy in babies is

something that parents simply want to see. Some are flattered, seeing it as a sign of their baby's love. Research in the United States found that when a baby is upset, parents appraise the situation, decide whether the baby is angry or jealous and come down much harder on anger than on jealousy.

First babies may get upset when their mothers make a fuss of another child, but this is something which confronts them far more regularly after a younger sibling is born. Even when parents use clever tactics like giving the child a present from the newborn baby, assigning them a helping role and making any changes to their routine well before the baby is born, some jealousy is inevitable. When he was a toddler my partner would deliberately wake his baby sister up while she was in the garden sleeping in her pram so that he could run inside calling 'baby wake' to his mother and receive praise for being so helpful. It wasn't until his mother saw him shaking the pram one day that she realised these weren't chance awakenings. I also knew a boy who, when his mother was standing on a hill chatting to a friend, nudged the pram with his foot, just enough to cause it to roll off down the hill towards the main road. Luckily the mother reached it in time.

Changes in behaviour are very common after the birth of a younger sibling; one study found that 92% of children began behaving badly. Even monkeys take on depressive postures after the birth of a new brother or sister. Not all children manifest their jealousy in aggression towards the baby; some become withdrawn and others regress – no longer sleeping through the night, demanding a bottle, talking like babies and forgetting their toilet training. Intriguingly Judy Dunn, a world authority in the area of sibling research, found that the clearest signs of jealousy were shown not when the mother tended to the new baby, but when the father or a grandparent showed interest. Many toddlers have been well prepared by their parents for the fact that the mother will need to spend a lot of time tending to the new baby, but the fact that other relatives are also interested in the new arrival can come as a shock. Those who had close relationships with their fathers before the birth showed fewer signs of jealousy.

There is contradictory evidence regarding the difference made by the sex of the siblings or the age gap between them. Some studies have found that children are more friendly to younger siblings of the same sex and more aggressive to the opposite, while others found the converse. Similarly with age, some have found that a three-year age gap can present most problems while others suggest that the larger the age gap the better, because the child is less likely to consider the baby as a rival.

Not all sibling conflict is a problem, however, provided the children also have plenty of positive interactions with each other, but Ayala Pines did find that sibling rivalry as a child could lead to sexual jealousy in adulthood. Strangely, she also found that the more older brothers people had, the more jealous they were as adults, while older sisters made no difference. Even when people don't become sexually jealous, rivalry with their sibling can continue into adulthood, even resulting in life-long competition. Occasionally this can go very wrong.

* * *

Harman and Vikram were students who shared a room together in London. They had a lot in common; both were very clever, both wanted to be doctors and they had the same parents. Vikram, the younger of the two brothers, did spectacularly well in his exams at sixteen and, like his brother before him, appeared to have a successful future ahead of him. It was in his very last year of school that things started to go wrong and he developed a drug problem, resulting in three convictions for the possession of heroin. He managed to give up the drug, but his A level results were disappointing and definitely would not win him a place at medical school. Instead he started a four-year biology course at university with the intention of training for an additional three years afterwards in order to qualify as a doctor. Meanwhile he watched his older brother Harman begin his medical training. He couldn't bear it. His brother had the life he wanted. Why should he have it so easy? By the time he had finished his biology degree Harman, always the golden boy, would already be practising as doctor, or he would be if Vikram didn't put a stop to it.

On the morning of Harman's medical exam he and Vikram were once again arguing over whether Vikram's university was any good – just a polytechnic, Harman said, leaving the house to set off for his exam. Minutes later Vikram left to go to college, but he couldn't stop thinking about what Harman had said. How dare he run down his college? What was so good about his fancy medical school anyway? He sat on the tube wishing he could have his brother's easy success and then had an idea. Searching in his bag for a scrap of paper, he began writing a script, a script for the new part he was about to play. Minutes later he emerged from the tube station, found a phone box and dialled 999. Police, please. He told them he had just overheard a man with an Arabic accent discussing five bombs planted in the very college where his brother was taking his exam.

Two weeks earlier terrorist bombs in Madrid had killed 190 people and the police were on constant terrorist alert. They acted at once, dispatching three bomb disposal teams to the college to begin a search. But for Vikram, everything did not go to plan. The building was not evacuated. The exam was not even disrupted. And twenty minutes after making the call Vikram was arrested outside the phone box from which he had made the hoax call. Harman took his exam as planned.

As a result of this one act of jealousy, not only will his brother qualify first, but he will probably be the only doctor in the family. After a six-month jail sentence, the likelihood of Vikram becoming a doctor seems further away than ever.

jealousy v. envy

Strictly speaking Vikram was suffering from envy rather than jealousy. Jealousy involves three people and the fear of loss of one of those people to another, while envy more often involves two people, one of whom wants something the other has, whether it be a possession, ability or success. While jealousy is often used in everyday conversation to refer to envy, the same does not happen the other way around. You wouldn't hear someone say, 'My wife's flirting with that man over there and maybe having an affair with him. That makes me envious.'

Despite the differences, Peter Salovey found that envy does bring up the same sorts of feelings as sexual jealousy, but in a milder form. He suggests that because three people are involved in sexual jealousy the feelings are that much stronger, but while jealousy involves feelings of rejection, hostility, anger, fear of loss, hurt and suspiciousness, envy brings in feelings of inferiority, dissatisfaction, longing and self-criticism.

Envy can still result in extreme behaviour, even in the strangest of circumstances. In the chapters on fear and disgust I mentioned Primo Levi's experiences in Auschwitz. He also described the distinct envy he felt when new people arrived at the concentration camp. Surprisingly it was not the case that new prisoners were always befriended and shown the ropes by those who were accustomed to the rules of the camp. 'The newcomer was envied because he still seemed to have on him the smell of his home. It was an absurd envy, because in fact one suffered much more during the first days of imprisonment than later on when habituation on the one hand and experience on the other made it possible to build oneself a shelter. He was derided and subjected to cruel pranks, as happens in all communities with "conscripts" and "rookies".'

In these extreme circumstances envy made an already appalling situation even worse for the new arrivals, but on a smaller scale envy also encroaches into everyday life. The economist Daniel Zizzo found that people would happily deprive others of money even if it meant losing money themselves. In the laboratory four people at a time gambled on which numbers they thought would come up on a computer screen, winning cash when they guessed correctly. Although each person was in a separate cubicle, unable to see what the other three were doing, the screen kept them updated on how much everyone had won and while everybody won something, one might have £20 while another had only £5. At the end of the game, they were told that they could choose to reduce other people's winnings, but only if they were prepared to sacrifice some of their own cash. They called it 'burning other people's money' and the cost of doing so was varied across the experiments, but it could cost as much as 25p to destroy a pound that someone else had won. It was stressed that this was just an option; they were free simply to walk away with the money they had won. The results were extraordinary: even at the highest price 62% of people were happy to pay some of their own money in order to stop someone else having so much. There was no gain for them – it was pure dog-in-the-manger stuff. Less surprisingly the people who had the most money taken away from them were the ones who had won the most. In many groups every person ended up worse off, when they could all have left the laboratory with more money. It was all down to envy or so it seemed.

There is a problem with this experiment. Everyone playing knew that the burning money option was also being offered to the others in the game, so they might have feared that everyone else would burn their money, in which case they would want to burn money too in a kind of pre-emptive revenge. They had to punish others in case they punished them. The experimenters must have been delighted as they watched their participants leaving the money behind, making their experiment cheaper by the day.

Perhaps this is why we take such care not to induce envy in others. If somebody admires a beautiful new house, the owner is quick to point out the downsides of the area. The American economist Robert Frank told me that he was once offered a red Porsche convertible at a bargain price after a relative had bought it in France and then discovered he was unable to register the car in California. Robert Frank lived in New York State where the car registration wouldn't present a problem, so he was offered the bargain of a lifetime. In the end he and his wife decided not to buy the car because they wouldn't be able to explain to people that they had bought it at a discount and were embarrassed to look as though they could afford a shiny, brand new Porsche.

It must be said, however, that envy, like jealousy, can have its uses. If an inequity is spotted, people can fight for equality of treatment, for example in civil rights campaigns. On an individual level envy can spur people on to work for the success they see in others. The importance of role models for young people is often emphasised. Perhaps the reason those role models are important is that while the people can identify with them, they also inspire just enough envy to encourage others to follow the same path.

Looking at jealousy and envy as a whole, perhaps the situation is similar to fear. We need to be slightly oversensitive, because it's better to be wrong occasionally than to ignore a serious threat to your relationship or, in the case of envy, an opportunity to change things. However unpleasant it feels to suspect that your partner might be interested in someone else, that early warning signal might give you the chance to hold onto them. This is preferable to not having an inkling until it's too late and they have left. As with all the emotions the question is how to find the balance between too much and too little. Philemon and Peribanez, divided by four centuries of play-writing, both had too much. Peribanez trusts his wife until the crucial moment where he sees a beautiful painting of her and realises that it was commissioned by a man far more powerful than he. From that instant, the jealous rage Peribanez experiences reveals that death is inevitable.

> 'Backwards, backwards, time unravel. Please not my mind, oh God. Don Fadrique has a painting of my wife. Just that leaves me sliced open to the vultures. But if she

didn't know – does she not know? Maybe nobody knows. Maybe nobody knows. This is the cost of her beauty. Is the price of peace to marry a wife you don't love? He's robbed me of my peace, dear God. Casilda, my Queen. This is so ugly, this jealousy. I don't want it to be seen, I don't want anyone to know what he wants, what he wants to do with her. It's like he's had her already. Chewed her up and spat out. I have to stop . . . I can't creep quietly back to my harm and home amidst whispers and gossip – everything I loved becomes my enemy. I need to talk to Casilda. Is she more than I deserve? What a stupid man I am. How could I think she could have been mine? That she was mine? How stupid to think that powerful wealth wouldn't want her too. Flick he envious eyes over her sweet face. Where would he flick his tongue? This will kill me. If this is what paint does, the real thing will send me to Hell. Help me Saint Roque. I can't live with this. God protect me I'll kill him. Stupid stupid why was I so stupid?'

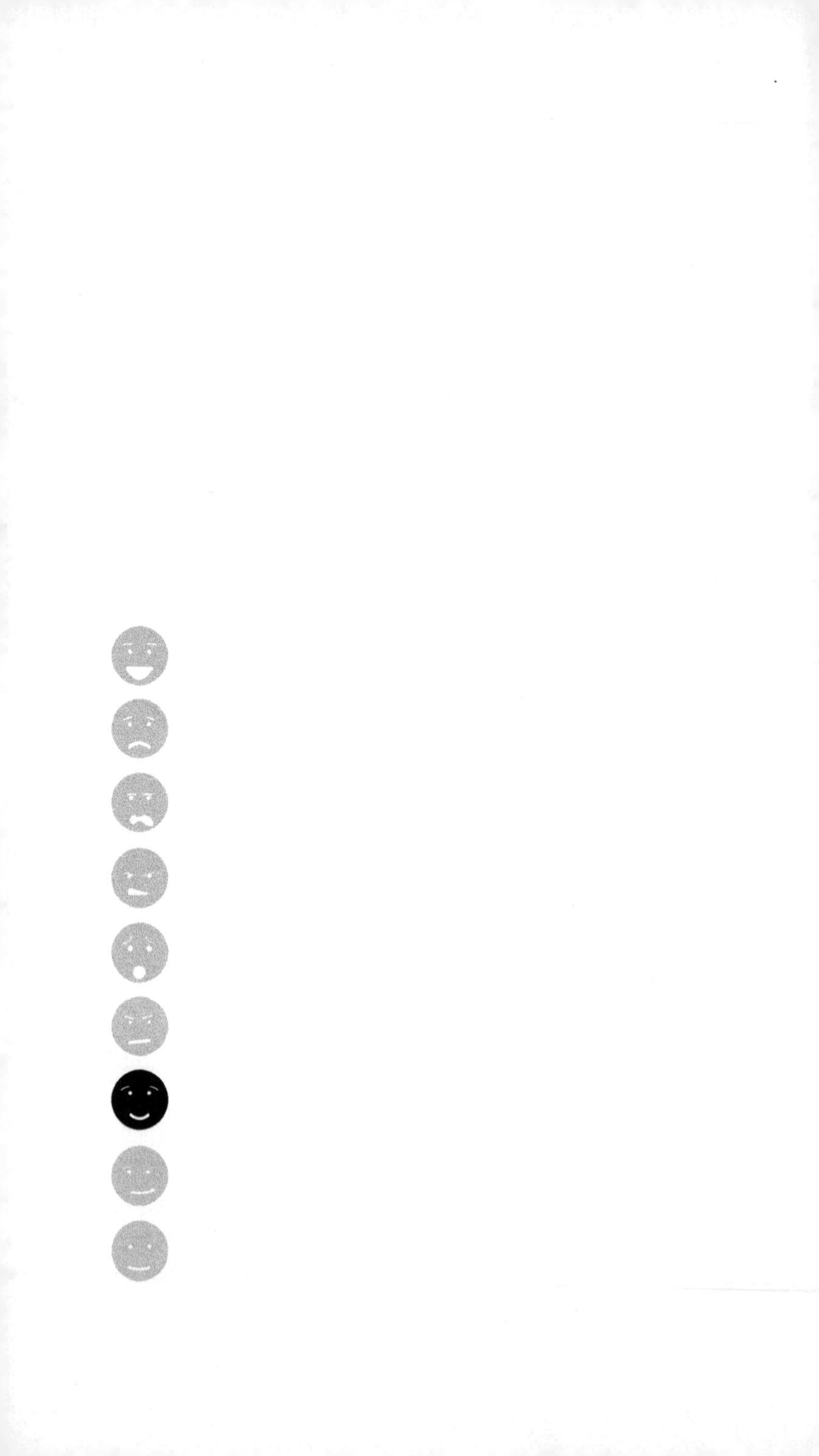

seven

love

One at a time each man and woman is led down into the pitch black restaurant by a waiter wearing night-vision goggles. It is so dark that you can't even see your own hand, let alone what you're eating. 'This is your seat. I'm putting your right hand on your wine glass and your left on your bread roll. Here are your knife and fork. Don't try pouring wine yourself. We'll come round and do that. If you want to go to the loo put your hand up and a waiter will come and guide you. Don't think your eyes will get used to the dark. They won't. See you later. Enjoy yourself!' There are eight people on this table. The four women are all wearing new tops, immaculately-applied blusher, plenty of mascara and of course Touche Eclat to cover up any bags under the eyes. These women know how to make themselves look good, but it's all in vain. No one can see them. Everyone giggles nervously, laughing at the difficulty of finding their wine glass in the dark. They introduce themselves to each other and start chatting. Soon they are giving their views on Hollywood and Brazilians – not something they would normally discuss with strangers, but this is dating in the dark where no one behaves in quite the usual way. With their night-vision goggles only the waiters are all-seeing. They notice that the statuesque blonde with incredible cheekbones is getting on fabulously with the short man with dough-like features. There's a lot of touching going on. Save shouting 'Oy, you opposite, I'm talking to you,' it's the only way of getting people's attention.

This is yet another dating gimmick, but it certainly is different. Soon people start describing their own looks to each other; it's not something any of them can ignore for long. After dessert all is revealed, but only gradually. Candles gently introduce each person to the sight opposite. Everyone braces themselves, ready to conceal their disappointment if necessary, although in the end everyone simply laughs. Afterwards some people insist that they did find themselves attracted to people they would not usually have considered, although no unlikely-looking couples appear to leave together.

The pitch black room is an attempt to let people get to know each other before looks can get in the way. It's the opposite of speed-dating where all you can decide in three minutes is whether the person opposite you is attractive and whether or not they are annoying. At vast organised dating evenings a lot goes on looks. A Polaroid photo is taken of each person as they arrive and placed on a board, accompanied by a little pouch. If you like the look of a person's photo you can put your card in their pouch. By the end of the evening some of the men's pouches are crammed so full that they swing precariously from one drawing pin. Others have none. The world of dating is harsh and some are far more expert than others, with escape calls to a mobile phone all set up in case a blind date is so awful that it would waste an evening. One glance at a dating website with a friend who's searching confirms that looks are taken very seriously. 'No – too short. No – look at his hair. This one's nice. Met up with him at a bar last week and thought it went quite well, but when I logged on straight afterwards to see whether he liked me or whether he'd gone back online to look for someone else, he was logged on, so clearly he doesn't like me. No – sounds creepy. No – obviously thinks he's hilarious. There's this one though – a designer. I've been emailing him for a while, but not sure that it's really worth spending a whole evening with him. I phoned this man, but there's no picture. He's put "very attractive", but everyone does. If they just say "attractive" that means they're probably ugly. Everyone exaggerates. When I spoke to him on the phone I asked him what he looked like. He said he was six foot tall and well-built – in other words he's bald. People always mention their hair colour when they describe themselves; if they don't it's because there isn't any hair to discuss.'

It's not surprising that so much attention is paid to looks on dating websites; attraction is after all the first stage of love. Physically attractive people don't only get the chance to take their pick on dating websites, but from babyhood onwards they are treated better by others and are assumed to be more sociable, interesting and humorous. Elaine Hatfield who has spent decades researching love showed people photographs of various women and asked them to describe their characters. Apart from the occasional fault, such as vanity, the better-looking people were ascribed all the virtues. We even make judgements based on how attractive we believe a person to be, never having actually seen them. In a study which purported to be on the subject of getting to know people, the psychologist Mark Snyder from the University of Minnesota gave men a photograph either of a beautiful woman or a woman who was 'homely' (whatever that might mean). As predicted the men expected the good-looking woman to possess all the usual virtues, but this time the experiment went a step further. The men were asked to phone the women for a chat, but in fact the people who answered the phones weren't the actual women in the photographs. And when the men spoke to them something extraordinary happened. The men's expectations of the women's beauty affected not only the way the men thought of the women, but the way the women actually behaved themselves. The women knew nothing of the photographs, but when the men believed they were talking to someone beautiful, the women became more animated, confident and socially skilled. Meanwhile the women treated as unattractive started to fulfil that prophecy, becoming withdrawn and awkward. The conversations were taped and played to other people who then judged what they thought the women might be like. Those who had been talked to as though they were attractive were rated as more good-looking. So people are right to exaggerate their looks in dating ads. Not only are the people who say they are highly attractive more likely to be chosen, but when they finally speak on the phone they will come over better, simply because the other person believes them to be attractive. It seems that there is no room for modesty.

If a person can convince potential suitors that they are attractive this can bring another problem. Somehow the offers have to be narrowed down to find the people most likely to provide that special emotion – falling in love. One, admittedly glamorous, friend of mine has had more than 7,000 emails from men who have seen her on a dating website and want to meet up. She has developed her own methods of cutting down the options; anyone who doesn't make a joke is out, but so are men who use certain clichés such as 'I work hard and play hard', 'I like cosy nights in and fun nights out' and, apparently a very common one, 'Phew, you're so good-looking I've just picked myself up off the floor'. Also out is anyone who can't spell or who includes a photo where they have cut their ex-girlfriend off the side of the picture. The problem is that expectations are inevitably high. Given pages of men to choose from, why not look for the one that's perfect and discount the others forever on the basis that they don't have the eye colour you would ideally choose and tennis isn't really your thing. After just one or two dates friends of mine have finished with people for crimes as small as bringing them a single rose (too naff), wearing an oatmeal jumper (too dull), suggesting a weekend away together in Bruges (lacking in romance), standing with bad posture while stirring a white sauce (unacceptable behaviour) and buying tickets for *Rod Stewart – The Musical* (no explanation needed). They are not alone. Research has shown that people are harshest on potential partners when they know they don't have to see them again. In another of

Elaine Hatfield's experiments students agreed to go on a series of blind dates over the following five weeks, but half were warned that they would see the same date once a week for the duration of the study, while others would see someone new every week. Then they were told the names of their dates and shown videos of all the dates taking part in a group discussion. Afterwards they had to rate the people they'd viewed and all the students liked their own date the best, while those who knew they would be seeing the same person for five weeks liked them even better. It seems heartening that people were so keen to see the best in their dates, but Hatfield interprets this as attraction towards those on whom they knew they were to depend for company.

This tendency to have high hopes of those you don't know might sway people's reactions while they are dating. If you met someone through an advert, had a nice evening with them, but weren't head over heels in love with them, you could either meet up with them again or return to the adverts for another huge pool of people, any of whom might be fantastic. Your real date has to compete with an imaginary crowd of potentially ideal people. You could see them again two days later, but what if one of these others is amazing? The decision is made quickly. Even slow-dating becomes speed-dating.

But it's all about looking for love, the emotion about which we probably have the firmest ideas. Our ideal partner begins to form in our heads long before we reach our teens. By watching both their own parents and other people's, young children develop ideas about what love is and how relationships should or could be.

is love a basic emotion?

Love is rarely included in the lists of basic emotions devised by psychology researchers. It's not the sort of emotion you tend to feel at one moment which has gone at the next. You might feel more love for your partner when you happen to be on holiday in a stunning place, but you don't feel it just for a moment like anger, nor would you feel it without explanation in the same way as sadness. You wouldn't say, 'I just feel a bit in love today – can't explain why.' Although love can comprise several different feelings – you might begin with elation, hope and perhaps a slight fear or even a sense of helplessness, and then at other stages pride, anger and even hate – one study found that the two emotions most commonly mentioned in conversations between married couples were love and, intriguingly, regret.

You can look at love in a number of ways. Is it a feeling, an attitude you have towards somebody or a set of behaviours? James Averill, whose work on anger was discussed earlier, is wonderfully unromantic when it comes to love. He believes that saying you are in love is simply a shorthand for a list of ways in which you hope the other person will behave. So if you tell someone you love them and they respond similarly, this is a quick way of making agreements regarding fidelity, the way you behave towards each other on an everyday basis, and your joint hopes for the future. If these expectations are fulfilled you can declare yourself to be in love. He says we interpret our behaviour, compare it with the romantic ideal we have in our head and then come to certain conclusions. If we enjoy spending time with a person, miss them when they're not there, think about them a lot and hope to spend the future with them, we decide we are in love. We tend to think that emotions happen to us, but according to Averill they are just things that we do, ways of behaving. He believes the reason we fall in love is that society neglects the individual. Society can't love us so we find another way to receive love and by idealising the person we love we preserve our own self-esteem.

Averill is not alone in suggesting that love is simply created by society. Depending on what you read, romantic love was invented by twelfth-century French troubadours, by Dante or even Shakespeare. However, earlier examples of romantic love aren't hard to find. The most often-quoted is Sappho's sixth-century poetry. There was a time when it was assumed that romantic love was a western European invention which only existed in cultures without arranged marriages. This is not the case at all. In 1992 two American anthropologists William Jankowiak and Edward Fischer conducted an ambitious project looking at anthropological studies from 166 different societies around the world to see whether love always featured. The only problem was that these studies had been carried out by different ethnographers using various definitions of love, making it hard to tell whether references to love were euphemisms for sex or references to romantic love. In addition Jankowiak and Fischer searched folklore tales, love songs and stories of elopement and found that romantic love was evident in at least 88% of the societies they examined. They deliberately employed strict criteria in their definitions of love and wherever there was an ambiguous motivation for a relationship, it was omitted from the figures, so in fact the figure of 88% is probably conservative. In only one society had the anthropologist stated that there was definitely no evidence of romantic love. This suggests that love is far from a Western invention. Instead it is the emphasis placed on romantic love which varies from culture to culture. There's an assumption in romantic comedy films in the West that the audience want the leading man and woman to end up together, despite their perfectly nice existing partners, who are inevitably dumped. True love is perceived to be more important than loyalty to a current partner, which is not the case in every culture. Moreover in the United States the emphasis on marrying for 'true love' is increasing; in the 1960s when people were asked whether they would marry the perfect man – intelligent, good-looking, kind, witty and entertaining – but with whom they felt no chemistry, women often said they would. Today only 9% say yes. The rest say they would not marry unless they were in love.

is it possible to study the science of love?

'I believe that 200 million other Americans want to leave some things in life a mystery, and right at the top of things we don't want to know is why a man falls in love with a woman and vice versa.' This was how a United States Senator, William Proxmire, attacked the study of the science of love in 1975. This didn't stop researchers from continuing to study the subject and thousands of journal papers have been published since. However, with this particular emotion there is always that nagging question of whether it will ever be possible to understand love and whether the attempt might somehow destroy the magic of love. In one sense this is strange. Nobody suggests that the mysteries of sadness will be ruined by researching the antecedents of depression, but then sadness tends to be seen as negative and thus worthy of study, if only in an attempt to expunge it. However, love can also cause immense pain. Even if you consider love to be primarily a positive emotion, this still isn't a good reason for ignoring it when it comes to research. The chemical changes in the brain caused by feelings of ecstasy are studied, without any suggestion that happiness should remain a mystery. For some reason love is considered to be a special emotion unlike any other.

Whether the study of love can make it any easier to find and hold onto is a different question, but it seems to have worked for Elaine Hatfield whose research I've already mentioned. She lives in Hawaii with her husband whom she adores, and together they write about love. She examines the psychology of love and he, the history, but when she started out it was her research which so incensed Senator Proxmire. He ridiculed her by awarding her a Golden Fleece award to show just how absurd he considered her research to be. Even her mother's Catholic bishop wrote to her saying that there was no need for her to study love because for centuries the church has known all there is to know about love and sex. Another professor of love, Ellen Berscheid, estimates that the publicity and controversy surrounding her choice of research topic wasted two years of her personal and professional life. However, she and Elaine Hatfield both persevered and since then many journals have been launched and masses of research has been conducted into love and sex.

Since Elaine Hatfield and her husband have been happily studying love together since the 1980s I asked her whether everyone expects the two of them to have a perfect relationship, anticipating an answer of the 'like all couples we fall out now and then' variety. Instead she took me by surprise when she said, without irony, 'Yes, people do expect us to have a perfect relationship, but it is pretty perfect, so that's fine.' If their experience is anything to go by, perhaps researching love can make a difference after all.

what is love for?

So what can the science of love reveal? The secrets of attraction perhaps and the effects of love on the body and the brain, but first there is the more basic topic of the purpose of love. The finding that people all over the world, today and in the past, have fallen in love suggests that it is an emotion which is either a by-product of evolution or which we have evolved to experience because it offers an improved chance of survival and reproduction. Such an evolutionary approach would suggest that love is all about the encouragement of procreation, except that strong sexual attraction with no love involved whatsoever can take care of humankind's need to reproduce. Why develop these strong bonds and risk the torture of rejection? One possibility is that the offspring were more likely to survive with two parents to protect them and it was feelings of love which might have kept their parents together. David Buss, whose ideas about jealousy I discussed in some detail, believes that men and women approach love differently because they have different goals – supposedly women look for a man who is rich enough to provide for her children, while men choose young, attractive women to increase their chances of finding a fertile partner. Buss says this can lead women to pretend to be younger than they are in order to attract a mate. It's ironic that I have female friends of about thirty-nine or forty who do sometimes lie about their age, particularly if they're going on blind dates. But they don't do this in order to appear more fertile. In fact it's the opposite; they pretend to be thirty-three because they know that many men will run a mile rather than meet a woman of thirty-nine in case she's on the look-out for good father-material.

However it is always easy to find individual examples from certain cultures at certain times in history that can contradict evolutionary theories. In Britain today some people choose not to have children and it's hard to see how this can fit in with the overall pattern of evolution. Evolution should not be used to justify or explain individual behaviours, nor does it rule out individuality. Part of the beauty of evolution is the way in which a species constantly throws up differences. The problem is that those who favour evolutionary explanations for behaviour often fail to specify the time in human history when these behaviours best suited life, nor why we wouldn't have evolved at all in the thousands of years since then. Men's instinct is supposedly to attempt to fertilise as many women as possible, but even though there is less societal pressure than ever in the West today to spend your life with one person many men still choose to do so. In fact multiple one night stands are not even a very efficient way for evolution to achieve maximum fertilisations. A man does not know whether a woman is fertile on the particular day he seduces her and even if by chance she is, he only has a 20% chance of getting her pregnant on that one occasion. An alternative approach is to invest his time in one woman, guarding her from other men and giving him more opportunities to try to fertilise her. He would need to sleep with a huge number of women to give him the same chance of making a baby that he has with one permanent partner. So perhaps this is why his emotions have evolved to encourage him to fall madly in love with one person and stay with her.

As well as helping us to fulfil our biological needs, love also satisfies distinct psychological requirements. It provides us with human companionship and makes us feels valued and loved, whilst giving us the chance to care for someone else in the same way. If we fall in love we have one person who genuinely cares about our well-being, cherishing us in the way that our parents did when we were young. No longer do we sleep alone in one dark corner of a house, but like a baby we gain comfort from sharing a space with someone who loves us.

The kind of love that is in one sense the most moving and the most memorable is that which is unrequited. A favourite love story for a lot of people is Gabriel Garcia Marquez' *Love in the Time of Cholera* in which Florentino Ariza loves a schoolgirl named Fermina Daza, who lives on a Caribbean island. Initially she loves him too, but after his love letters are discovered she is sent on a long journey. By the time she returns she feels rather foolish for having loved him and later marries another man. Florentino, however, still loves her and decides he must wait for her husband to die and fifty-one years, nine months and four days later he is able to declare his love for her.

It is easy to see the benefits which love brings. In that sense it is a rational emotion, but with unrequited love it's harder to see a purpose and yet we know it can be the strongest love of all. It is clearly a waste of time and energy to long for a person whom you know to be uninterested in you. This sort of love is not going to bring you someone with whom to reproduce, nor someone who will cherish you, yet it still persists. Not surprisingly, studies show that people whose love is unrequited are more distressed than those whose love is reciprocated. However Arthur Aron, who has spent years studying love, believes unrequited love can be rational. Firstly people might overestimate their chances of success. Even if they appraise the situation realistically and realise the likelihood of forming a relationship to be slim, the potential benefits of success make it worth the gamble. Perhaps unrequited love can be enjoyable – better to have loved and lost and all that. Aron did find, however, that those who had experienced unrequited love were more anxious about relationships and felt less secure. Whether this led them to choose unattainable people or whether their past experiences of unrequited love caused them to feel anxious, we cannot know. For Florentino Ariza, however, his lifetime of longing and loving led him to see love in a new way, as something to be celebrated for its own sake, whether or not it was reciprocated. This he tried to explain to Fermina Daza in a letter, a letter he hoped would change her mind. 'It had to teach her to think of love as a state of grace: not the means to anything, but the alpha and the omega, an end in itself.'

the mysteries of sexual attraction

Love might well have a purpose, but why are we attracted to one person and not another? There have been attempts to break attraction down. One theory is that men find women with a 0.67 waist-to-hip ratio particularly attractive – in other words the waist is a lot smaller than the hips. As I discussed in the previous chapter attractiveness can be related to symmetry which in turn might be a marker of genetic health. Researchers Steve Gangestad and Randy Thornhill found that women tended to have affairs with men who were more symmetrical than their regular partners. Were they really looking for the healthiest genes? David Buss believes this is all part of a mixed mating strategy. Subconsciously you find someone who'll make a good life partner and meanwhile shop around for affairs with the men with the best genes. Of course it is possible that the reason lovers are more symmetrical is that when it comes to an affair people go more on looks (you haven't got to live with them so you might as well) whereas with the live-in partner women are looking for a different combination of attributes and will be more fussy in other areas.

As well as preferring symmetrical faces, we are attracted to people with faces like our own. At St Andrews University

David Perrett used a computer to morph photographs of women's faces into men's and vice versa. As a woman you go in, they take your photograph and later on you're shown a series of photos of men and asked to decide which is the most attractive. Among those men, although you won't realise it, is a male version of yourself and the chances are that this is the picture you will select as the most attractive. The possible reasons are that this a face which somehow seems familiar, that it's the face most like your father's or that by breeding with someone who looks like you, your own looks are more likely to be reflected in your offspring. It's often assumed that the most attractive people pair off together because they are all going for the best-looking person they can get, but perhaps the less attractive people are not settling for what they can get, but are attracted to someone who looks a little like them. The problem with all these studies is that choosing photos in a laboratory has little in common with real life. Everyone can describe their ideal physical partner, but often end up with someone very different. Hopefully a long-term partnership is based on a lot more than the colour of someone's hair.

the scent of love

Looks can't explain all the mysteries of attraction. Why is it that a person can look perfect on paper – good-looking, friendly, intelligent and kind – but when it comes to it, there's just something missing? In the eighties and nineties pheromones were the fashionable answer to this question. There were rumours that aftershaves containing scents from chimpanzees' gonads were enough to drive women wild. They were the mystery element of attraction. People would hear about nuns in convents and girls in boarding schools all having their periods at the same time due to the influence of pheromones. In fact far less research has been done on pheromones than their public profile might suggest. It is such a popular topic, however, that the few studies which have been conducted, often involving very small samples, generated huge amounts of publicity. In fact there is still disagreement over whether the chemical messages that humans might send to each other can actually be counted as pheromones in the sense that animals use them. As I mentioned in the chapter on fear, there are pheromones which signal fear in several animal species, but it's not clear whether this occurs in humans. Those who do study them, like Professor Karl Grammar from the Ludwig Bottzmann Institute for Urban Ethology in Vienna, believe that they should have been taken seriously sooner and that as a discipline psychology can be too quick to neglect unconscious processes. He proposes that when you meet somebody new pheromones could induce an emotion before you've had a chance to consciously process your thoughts about the person. The problem is that no one knows exactly what these substances are. Sweat contains more than 200 different substances and only a handful have been tested.

Those who study pheromones tend to do so as an adjunct to more traditional scientific work. Claus Wedekind, for example, from the University of Bern in Switzerland, is a zoologist. He studies fundamental questions in evolutionary biology, mainly using fish. When he was looking at mate choice in fish, he began using human subjects in his experiments for the simple reason that unlike fish, they can talk. In contrast to the way studies are usually carried out – experimenting with mice in the hope of discovering more about humans – he did the opposite. He wanted to know about fish, but humans were the most convenient subjects he could find.

His discoveries were extraordinary. A group of men were given clean white T-shirts to wear for two nights and instructed to refrain from wearing scented products during that time. Then each T-shirt was put in a box with a sniffing hole at the top. Lucky female volunteers were required to sniff each box in turn, rating each of seven T-shirts as pleasant or unpleasant. He found there was no one T-shirt preferred by all the women. In fact they did not agree at all. However, something extraordinary was influencing the women's choices. Each woman preferred the T-shirts belonging to men with different immune system genes from hers. Each person has different major histocompatibility complex (MHC) genes. The idea is that every immune system produces different bacteria on the skin which are then detected in sweat. In theory if you breed with someone with MHC genes very different from yours, you are unlikely to be related thus reducing the risk of inbreeding and passing on genetic diseases. However, Wedekind thinks the main reason for choosing someone with a different immune system from your own is to give your children a better combination of genes. This won't always work, because if women are on the pill it seems to mask the system, making them less likely to pick immune systems different from their own. However, MHC genes cannot provide the only basis for attraction. If they did, women would in theory cease fancying their partners as soon as they stopped taking the pill.

More recently Wedekind has found that we might even choose to wear perfumes which reflect our own immune systems. Rather than using scents to mask our own natural smell, he found that people tend to choose the perfumes closest to the smell produced by their own immune system. So we enhance our own smell, theoretically making it even easier for potential partners to detect our particular immune system. However, he did also find that although MHC genes do not show sex differences, men and women preferred different smells, so cultural factors clearly still play their part.

Karl Grammar even found he could confuse men using substances called copulins found in vaginal secretions. If men inhaled this smell they were no longer able to rate the attractiveness of a series of photos of women. Even though they couldn't consciously detect the smell, all the women became equally attractive. Unfortunately copulins disintegrate very quickly so currently they cannot be bottled and used to fool men into thinking you're beautiful. Grammar believes that unbeknown to us chemical warfare is occurring between the sexes. While women give off chemicals which confuse men's aesthetic judgements, men are releasing a pheromone which is unattractive to women unless they are ovulating. The idea is that this helps men to filter out the women who are not fertile.

The trouble is that, unlike Napoleon's famous message to Josephine, 'I'll be home in three weeks. Don't wash,' most of us do our best not to smell of sweat. Professor Grammar is concerned that the modern obsession with cleanliness might even disrupt our selection of good partners. However, there are a few things to remember before pheromones are declared the answer to everlasting love. These studies exposed people to higher concentrations of substance and for longer periods than would happen in reality. Pheromones can only make a difference if they are detected. If you find somebody physically repulsive you are probably not going to get close enough for their smell to suddenly sweep you off your feet. When it comes to sexual attraction there has to be more to it than pheromones. Everyone has known an initial attraction disappear within seconds of a person opening their mouth to speak.

when do we fall in love?

It had not been a good year for Christine. She had helped to nurse one of her oldest friends while he died from motor neurone disease. She had lost her job, fallen out with her sister and the attempts she made at dating all came to nothing. To make matters worse she had just turned forty. Here she was without a partner or a job, uncertain what she should do next. Meanwhile Martin, the younger brother of Christine's best friend from school, was equally unhappy. He had just been through a difficult divorce and was trying to sell the family home while looking for somewhere new to live where his eight-year-old son would be able to visit. Years before, at the age of eighteen, he'd been on holiday with his sister and had developed a crush on the older and wiser Christine. For Christine's part she had never thought of him in that way; he was simply her friend's little brother. The boys she was interested in were older and had cars.

Then came a big meal to celebrate Martin's sister's fortieth birthday. Naturally, as her best friend from school Christine was invited, but since she wouldn't know many people at the party the logical place to seat her was next to Martin. The attraction was instant. Christine had known him for such a long time that she felt an immediate closeness and he felt the same way. Soon they fell in love and were living together. There seemed to be something about meeting at that moment when they were both at an all-time low which led them to fall in love. Christine says that the previous year had been so bad that she particularly appreciated any nice time she spent with Martin. 'Just feeling so comfortable together was enough. We've talked about this and maybe at other times we'd have both looked for more, but we both wanted to meet somebody and it just felt right. Twenty years ago I'd never have guessed that I would end up so in love with my friend's little brother.'

Back in the 1940s it was suggested that people are more likely to fall in love when their self-esteem is low, which would fit in with the experience of Christine and Martin. In 1965 Elaine Hatfield put this theory to the test. She wanted to know whether people are particularly receptive to affection after their confidence has taken a bashing. She gave female students personality inventories and then faked the results, leading them to believe that they were either better, worse or roughly as they might expect. No one would relish hearing the following about themselves: 'Although you have adopted certain superficial appearances of maturity to enable you to adjust temporarily to life situations, your basically immature drives remain.' They were criticised for having weak personalities, a lack of openness, an over-estimation of their own assets, anti-social motives, a lack of flexibility and a tendency to try to cover up their weak points in an attempt to impress others. Subjected to such a comprehensive character assassination one response is to decide that the personality test is at fault, but Elaine's hope was that instead of blaming the test their self-esteem would suffer a temporary dent. Meanwhile the women lucky enough to be in the group whose self-esteem was raised were told they were fantastic and that their personality profile was one of the best that they had ever seen. The third group were told their tests had not yet been marked. Then came the entertaining part. Each student is left in a room alone and in walks a handsome male postgraduate student. They chat and he invites her out for dinner and a film the following weekend. Later the women were asked to give their first impressions of this man. Those whose self-esteem had taken a knocking found him the most attractive. The theory is that they felt most in need of appreciation from someone else, while the women whose self-esteem was now sky-high felt they deserved someone better than this man and were less impressed. This could explain why people sometimes fall in love on the 'rebound', after their ex-partner ends their relationship.

Whether or not low self-esteem helps people to fall in love, once it has happened their self-esteem does go up. Arthur Aron followed a group of more than 300 students during the autumn term at college to see whether they fell in love. Ten weeks might not seem long, but he knew it was a time when students would meet lots of new people, so the potential for love was there. He was correct in his assumption because almost a third did claim to have fallen in love during this time. There were a few problems, such as students not turning up for every part of the study and the loss of one set of data, but it's all in a day's work for a research psychologist and Aron did manage to show that self-esteem goes up after people have fallen in love, regardless of a person's mood on the day they were tested.

We base our ideas of romantic love on the love we experienced from our parents which means we are more likely to fall in love in situations which recreate our childhood. Many of the activities new couples do together like going on daytrips, running a nice bath for the other person or surprising them with a present, echo the parent–child relationship.

There are four situations in which people are particularly likely to fall in love: when they feel lonely, dissatisfied, in need of sex or in need of variety. According to research a person needs something they can interpret as interest from the another person plus half a day to think about them. After that they're well on their way to falling in love. Perhaps it's not surprising that the biggest predictor of whether a marriage lasts is whether either partner happens to meet someone new at the moment when they're feeling dissatisfied with the marriage. If they meet someone while they're satisfied with their relationship they tend to stay married. Likewise if they are unhappy, but don't happen to meet someone else they will often stick it out with their partner until things get better.

love on a bridge

The Capillano suspension bridge in Vancouver sways 250 feet above a gorge that is almost twice as wide as it is high. Tourists pick their way across, clinging to the ropes and leaning to and fro to sway the bridge. In their minds are visions of enemies slicing through the supporting ropes with a Samurai sword until the bridge falls in a huge arc and ends up draped vertically down one cliff while people hang onto

the tip. Thoughts regarding this sort of improbable scenario were crucial to a famous experiment conducted on the bridge in 1974. Single men crossing the bridge were met by a pretty woman in the middle who asked them to take part in a study. She showed them an ambiguous picture, purporting to be part of a study examining the influence of scenery on creativity. After the men had told her a story to fit the picture, she gave them her number in case they wanted to contact her later for a copy of the results. She also conducted the identical study on a safe, solid, cement bridge further downstream, but the results were rather different. The men she met on the suspension bridge used more sexual imagery in their stories and nine out of thirty-three phoned her up afterwards. Of the men on the safer bridge only two called her. This is interpreted to mean that the men on the scary bridge found the woman more attractive. Something about the danger made them like her more.

Donald Dutton and Arthur Aron, who conducted the study, believed that in fact these men were simply misinterpreting what they felt in their bodies. While they were on the first bridge the men felt their heartbeat quicken and without realising it, looked around for a reason and decided they must fancy the woman. However, this a rather strange explanation. This was not an ambiguous situation. They were all too aware they were on a precarious bridge, so when they became aware of their fast heartbeat the cause should have been easy to spot. There is an alternative explanation. We know that when any unpleasant stimulus stops we feel pleasure. Take as an example, the huge joy and relief experienced by the people I met bungee jumping in New

Zealand. No wonder the men on the bridge experiment liked the woman. They are walking across a bridge feeling afraid and then they meet a friendly woman who distracts them and stops them being so scared. Maybe they were feeling relief. There is a also a leap in logic in assuming that all those who phoned her up afterwards had only one thing in mind. They didn't actually proposition her. They just phoned up for the results. Call me old-fashioned, but couldn't we credit at least some of the men with actually being interested in the results? No one ever asked the men whether they had any feelings for her.

In the years following the love on a bridge experiment researchers tried other tactics to see whether attraction can be induced. In one study men had to run on the spot for either 15 or 120 seconds. Then they were either shown a video of a woman who was dressed attractively or as a 'slob'. The men who had jogged for longer, whose heart rates would be higher, found the nicely-dressed woman more attractive than those who only jogged for a short period did. So perhaps they were misattributing their symptoms and believed they were attracted to the woman. However, the results of the woman in the 'slob' condition confuse things rather. Here it was the men who had only jogged briefly who found her the most attractive.

The idea that we can misinterpret symptoms, causing us to believe we are experiencing a particular emotion, wasn't just some strange idea dreamt up by the researchers conducting the love on the bridge experiment. They were following up one of the most famous experiments ever conducted on emotions. In 1962 Stanley Schachter and Jerome Singer developed a two-factor theory of emotion. They proposed that you had to both feel a physical sensation and label its cause in order to experience an emotion. To investigate this further they injected volunteers with epinephrine, but told them it was a new vitamin injection. One group was warned of the genuine side-effects of the drug such as shaking hands, flushed faces and a racing heart. A second group were told it had no side-effects and a third were deliberately misinformed that it caused headaches, itchiness and numb feet. One at a time each person was taken to a room and there along with another man taking part in the study they were given a questionnaire to complete. The other man, who was in fact a stooge, would then either start mucking about with a hula hoop and throwing screwed-up pieces of paper into the bin or he would become incensed with some of the personal items in the questionnaire, such as how often they had sex, which family members bathed regularly and how many lovers their mother had had. For the latter question the reply options started with 'four and under'. Eventually the angry man would become so annoyed that he would storm out of the room. The idea was that these moods would be infectious. It was predicted that people who were unaware of the side-effects of the drug would look around for an explanation for their physical sensations and assume it must be because they were feeling euphoric or angry and as a result they would feel these emotions more strongly. This is exactly what happened.

The trouble is that attempts by other researchers to repeat this experiment have failed. Other evidence suggests that you don't have to experience physiological symptoms in order to experience an emotion. However, Schachter and Singer did highlight something very important about emotions, which is that our view of a situation influences the way we experience certain feelings. So if the men on the bridge did mistake their fear for love, then we have to start considering love not simply as something that happens to us. Although we talk about 'falling in love' our thoughts play an active part in it. This is why we're more likely to fall in love at certain times than at others. Having said this, there does seem to be something special about falling in love in fearful situations.

In 1995 a thatcher named Simon was travelling across the Arctic ice-cap with five of his friends, all experienced climbers, with the aim of reaching some remote peaks in Greenland. They intended to be the first to climb them. It was snowing heavily and as they pushed on their cross-country skis they knew there was always the theoretical possibility of falling into a crevasse, hidden for centuries in the ice below, but it was unlikely and if it did happen they all knew the safety drill. They were roped together in groups of three so that if the person at the front fell through the snow, the other two could anchor themselves into the ground and set up a pulley system to haul them back up to the surface. Simon was in the middle of his threesome, with his friend in front and his friend's little sister, Sarah, behind. As he plodded onward romance was the last thing on his mind. They were making good progress and his friend was marching on ahead, when suddenly Simon felt his skis sinking into the snow, but there was no snow for him to sink into. The ground beneath him gave way and he fell down into a deep fissure in the ice, followed by his sledge which tumbled down onto his head, knocking him out. As he and the sledge plunged down into the crevasse Sarah, who was only slight, was almost pulled off her feet as she was dragged towards the edge of the drop. Remembering the drill she executed a manoeuvre known as an ice-axe arrest. Throwing herself to the ground she whipped her axe out of her harness, forced it into the snow and leant her shoulder onto the axe, putting her whole weight behind it. Simon was coming in and out of consciousness, hanging from the rope, with a thirty-foot drop onto the ice below him. His life depended on the ability of Sarah with her tiny frame to support the weight of both him and the sledge. If she couldn't hold him she would have to choose between falling over the edge herself or cutting the rope, which would probably kill Simon. She managed to cling on, while the other group caught up, set up the system of anchors and pulleys, and hoisted him back over the top. Simon's head was bloody, clearly injured by the fall but despite the mess he was still smiling, or he was at first. Soon he was having trouble maintaining consciousness and the doctor in the group knew they needed help and fast. He set off an emergency locator beacon and then they waited, aware that since they were in one of the most desolate and remote places on earth any help would take a long time. Three days later they were still waiting and had to accept that the beacon must have failed. So they let off a second beacon and the waiting began all over again. By chance a transatlantic jet was flying past at that moment, thousands of feet overhead and the pilot picked up the signal, radioed the nearest airport and an emergency helicopter set off to rescue the stranded group. Sarah had saved Simon's life.

Undeterred by the accident, Simon returned to the Arctic for four months the following year. When he arrived back at the harbour in the mining settlement of Longyearbyen, there was a small figure waiting for him on the quay. It was Sarah. Something about the terrifying situation they shared had drawn them together and today Simon is married to the woman who saved his life.

This shows that love can come out of fear. Unintentionally Aron returned to the subject of falling in love in fearful situations fifteen years after the love on a bridge experiment. He and his wife were conducting the study I mentioned earlier looking at whether self-esteem rises when people are in love. The day after the students had returned to have their self-esteem measured for the second time in the study, something happened which accidentally made fear an extra variable in the experiment – a big earthquake in Loma Prieta with an epicentre just ten miles from the university. This misfortune gave Aron the perfect opportunity to discover whether the fright of an earthquake left students more likely to fall in love. In fact the numbers of students finding new partners was the same as during earlier parts of the study but this time those who did fall in love were the students who had been the most distressed immediately after the earthquake. Perhaps it was love on a bridge all over again and they felt anxious; then a misinterpretation of those symptoms led them to assume they were falling in love. Alternatively, maybe they were more receptive to offers because they were seeking comfort after the quake.

love is good for you

Once you are in love the effects of the emotion are felt both in your brain and in the rest of your body. Everyone knows that love can feel good, but it might also bring health benefits. Janice Kiecolt-Glaser and Ronald Glaser, yet another married couple involved in love research, conducted an experiment on other pairs of spouses. They used a special vacuum to create a line of pea-sized blisters on the arm of each volunteer. Then each couple was asked to discuss together an aspect of themselves they would each like to change. Meanwhile the blisters were monitored to see how fast compounds called cytokines were delivered to the area to begin the healing process and blood samples were taken to measure the levels of the stress hormone cortisol. It was found that the couples' conversations could actually influence the way the blister healed; if the couples had a positive conversation their cortisol levels dropped and the blister healed faster, but extraordinarily negative discussion delayed the healing process. It is of course a big step to infer from this that love is affecting the immune system. Perhaps it was relaxation which made the difference. However, a study of Danish college students found that those who were in love were healthier and had unusually low activity in their natural killer cells (a good sign). Those whose love was unrequited suffered from more colds and sore throats and their immune systems showed signs of increased activity, implying attempts to fight off disease. Of course there is one problem with this study. The 'in love' group might have been healthier in the first place. Perhaps those who weren't in love were too busy coping with poor health to have met the partner of their dreams.

love hormones

In 1819 the writer Stendhal came up with the idea of writing a book on the science of love, informed by his experience of falling in love with a woman called Mathilde Viscontini Dembowski who sadly didn't feel the same way about him. In one of his many prefaces to the book he complains that in the first eleven years after publication only seventeen people read his book. In a later edition he starts the preface with 'This book has met with no success; it has been found unintelligible and not without cause'. However, Stendhal's book has since become a classic, with its explanation of the seven stages of love – from initial admiration to anticipation of what you could do together, to hope, to the enjoyment of seeing the object of your love through to the realisation that you are in love and then that first 'crystallisation' that your feelings are reciprocated. He doesn't stop there, which might be why his ideas eventually struck such a chord with people. In stage six doubt creeps in and you wonder whether they really do love you, but this is hopefully followed by the final stage, the second crystallisation where you realise that they do. These stages are not invariant. He says that in fact a person in love vacillates between considering the other person's perfection, thinking how wonderful it is that they love you, and wondering how to get the strongest possible proof of their love. Although Stendhal referred to his work as a book on the physiology of love there is little attempt to examine what is happening within the body. However, more than 180 years later, researchers are examining the chemicals associated with each stage of love.

Helen Fisher from Rutgers University in the United States has divided love into three stages, each of which has its own biological concomitants. For lust there's testosterone and oestrogen, for romantic love there are high levels of dopamine and norepinephrine and for a period of longer-term attachment there's vasopressin and oxytocin.

At the initial stage of attraction, as well as the lust fuelled by testosterone and oestrogen, other chemicals might also play a part, one of which is called phenylethylamine (PEA). It occurs naturally in the brain and acts like an amphetamine. The psychiatrist Michael Liebowitz believes that it's this substance which makes falling in love so joyous, and so addictive. He says the reason infatuation wears off after a time is that the brain can't cope with this elevated state for a long period and either the nerve endings stop responding to the chemical or levels of PEA drop. Could this be why love wanes? If someone is suddenly dumped could part of their misery be caused by withdrawal symptoms as the brain tries to cope with the drop in PEA? Liebowitz noticed he was treating many clients who craved relationships, picked unsuitable partners, were inevitably abandoned, experienced despair and then started all over again. He believed that the problem was in their brains, that they were in need of PEA highs. When he gave them anti-depressants which blocked the substances that break down PEA, thus leaving it to act in the brain for longer, things changed. One man even began to choose his partners more carefully.

PEA, along with neurotransmitters like dopamine and epinephrine, could account for the elation that love brings. Meanwhile neurotransmitters like serotonin which usually make you feel happy, drop to low levels when you're madly in love. Intriguingly they drop to the same low levels found in people with obsessive compulsive disorder, reflecting perhaps the obsessive nature of love.

As a love affair continues other chemicals gradually come into play in the brain, one of which is oxytocin, sometimes known as the love drug. Confusingly oxytocin acts both as a hormone (causing contractions during labour and the release of milk in lactating women) and as a neurotransmitter with distinct effects on the brain. Although it is produced in a small area of the brain just above the roof of the mouth, the neurons project into other parts of the brain and even affect the spinal cord.

When a mother has just given birth she is flooded with the hormone oxytocin. Even more is released while breastfeeding and many midwives believe that this makes the first hour after birth a crucial time for mothers and babies to bond. Rats start displaying maternal behaviour if oxytocin is injected straight into their brains, whereas if oxytocin in blocked they lose all interest in their young. The same phenomenon occurs in adult rats when it comes to pairing up. Give them oxytocin and a pair will become inseparable, block it and they cease to care. If pigeons are injected with oxytocin they start mounting and mating at once. But this substance also seems to influence human adult relationships. Just sharing a meal together can cause a couple to release oxytocin and during orgasm levels are three to five times as high as normal. It has been suggested that in women oxytocin causes the uterine contractions at orgasm which might help to waft sperm up to meet the egg. It is possible that the warm, comfortable feelings experienced after sex are provided by oxytocin.

the life-long devotion of the prairie vole

A small brown creature about the size of a mouse nestles in the long grass in front of a grave stone in a dusty cemetery in the American mid-west. A similar creature approaches and they start mating. These are prairie voles. Little do they know that they are of great interest to researchers looking at the biochemistry of love. The reason they are so special

is that they are one of only 3% of species of mammals which are monogamous and what makes them the perfect research tool is that they are similar to a darker vole found in the Rockies, called the montane vole, which is not monogamous and lives alone rather than in family groups. Between them these two species of mice provide researchers with the opportunity to study the brain circuits of love.

With fear it was easy. Teach rats to be afraid of something and then try to unravel the workings of their brains. Unfortunately you cannot train mammals to fall in love or even to form bonds with another animal. However, if you take a pair of devoted prairie voles, you can study their behaviour in different circumstances and then compare it with the way montane voles behave. Tom Insel found that when the production of oxytocin was blocked in female prairie voles, the bond with their mate was broken and they would mate with any vole. The same thing happened with males if they blocked the production of another hormone called vasopressin. The males also stopped protecting their mates from other voles. This process was absent from the montane voles, suggesting that it's the way the brains of prairie voles respond to oxytocin and vasopressin which causes them to live monogamously. Insel and his colleagues have discovered that receptors for oxytocin and vasopressin are distributed very differently in the brains of the two types of voles. He's even managed to breed a transgenic vole which is a montane vole, but has the gene for vasopressin reception normally found in a prairie vole. This vole's behaviour was transformed. He suddenly became more sociable. For the first time the alteration of one gene had been shown to have a dramatic effect on behaviour.

It is difficult to extrapolate from this research to human beings but in his future work Insel plans to examine the distribution of receptors for these two neurotransmitters in the brains of humans to see whether this could shed any light on why some people find it particularly difficult to relate to others.

Oxytocin has not been demonstrated to be essential to human attachment or romantic love. How many volunteers would there be for an experiment where you arrived as a couple and were given an injection which prompted you to abandon your partner, albeit temporarily, and have sex with someone else at the laboratory? On second thoughts, maybe there would be plenty.

Rebecca Turner from the University of California, San Francisco has succeeded in devising a way to study the effect of oxytocin on humans. She measured oxytocin levels in the bloodstream while women talked about past loves, happy or unhappy. It made no difference to oxytocin levels whether they were talking about positive or negative love stories, but the researchers did notice that in some people oxytocin levels shot up particularly high when they discussed love. Examining the results in more detail they found that the women whose levels rose highest had more secure relationships with fewer problems. Of course we don't know which came first. Were they born with a brain that is more likely to let oxytocin flow freely, which in turn helps them to form secure relationships, or is more oxytocin released as a result of the stable relationships they have experienced?

There is evidence to suggest that high levels of oxytocin make you feel calm. Women often describe breastfeeding as a dream-like state where nothing can disturb them and studies comparing mothers who breastfeed with those who bottle-feed confirm that breastfeeding mothers have lower levels of certain stress hormones. The problem here is that we don't know why they chose their method of feeding. The bottle-feeding mothers might have been more stressed in the first place, which could have contributed to their decision to use a bottle. However, panic attacks in women with anxiety disorders have been known to disappear while they are breastfeeding, only to return after the baby is weaned. There's a classic laboratory test of reactions to stress called the Trier Social Stress Test where the poor participants have to imagine they are at a job interview. They then spend five minutes explaining to an interview panel of three why they would be good at a particular job. Meanwhile the three people on the interview panel give them no feedback at all, not even a smile or a nod, and it's this which makes people anxious. Margaret Altemus from Cornell University used this test to see how breastfeeding women fared compared with other women and found that those who breastfed found the test less stressful. This suggests that the presence of oxytocin protects against stress.

Taking the evidence as a whole, it is possible that both oxytocin and vasopressin aid the attachment process between a couple. These hormones are very similar in terms of chemical make-up, but have very different effects on the body. Recently oxytocin has been implicated in feelings of love in new research coming from a rather different direction – work examining exactly where in the brain love might take place.

Are you deeply in love? We need your brain

This was the message on the posters seeking participants for a study by Andreas Bartels at University College London. Unfortunately the reaction to the poster was not as large as he had hoped, so he emailed thousands of students at the university instead. This time he received plenty of replies, three quarters of which came from women. Bartels used functional magnetic resonance imaging to scan the brains of people in love. Previous studies using brain scanners have tended to focus on negative emotions like fear or sadness. Although the technology has been around for more than twenty years, love has been left out. After Bartels had finished a PhD on the way the brain processes colour vision he was looking for a short project which might provide a contrast and the idea he came up with was love.

When his love-struck volunteers came to the lab they were interviewed and given a copy of Elaine Hatfield's Passionate Love Scale to assess just how committed to their lover they were. He only wanted the brains which were truly in love. You might expect these to be people in very new relationships, in the first flush of love, but in fact the average length of relationship was two-and-a-half years. Each devoted person was asked to provide photographs of their partner and of three friends of the same gender as their partner whom they had known for roughly the same length of time. Then they lay inside the narrow, noisy scanner –

not the ideal place for thoughts of romance – while they looked at each photo in turn and thought about the person pictured. The scanner provides a three-dimensional image of the brain every four seconds which can tell you the amount of blood flowing every 3mm throughout the brain. At any given moment you can see which areas of the brain are in use more or less than usual. Bartels was astonished to find such distinct differences between the brain areas people used to think about their lover and their friends.

Love led to activity in four areas buried deep within the older parts of the brain. Interestingly these were all areas which are also stimulated by cocaine. One was the *striatum*, an area involved in making you feel happy. Another, more intriguing area was the anterior *singular gyrus*. This is used for processes associated with putting yourself in someone else's position or seeing something from another's perspective. Bartels speculates that perhaps in this case it was related to the knowledge that you are loved by the other person. It would be interesting to see whether this lit up for unrequited love like Stendhal's.

Another area involved in love was the middle *insula*. This is a region which deals with the crossover between emotional reactions and physical feelings. Perhaps in the case of love a partner's touch is translated into a positive emotional feeling. It's an area involved with gut feelings which might account for that butterflies-in-the-stomach feeling associated with love.

The areas where activity was reduced are just as revealing, areas like the amygdala which is generally associated with negative emotions like fear and anger. Regions involved with understanding what other people might be feeling also ceased to be used, suggesting that once you're in love you lose the ability to assess the other person objectively.

Fascinating as these findings are, there is a problem. How do we know that love was the feeling people were experiencing while they were in the scanner? Faced with photos of your partner and three friends whom you like, there are two other differences between your lover and your friends. One is familiarity. You are likely to know your partner's face even better than those of your friends. Luckily this can be ruled out because we know which areas the brain uses to process familiar faces and they don't match up with the areas which lit up in this study. The second feeling might be more of a problem and that is lust. The chances are that you fancy your partner and not the others, so perhaps the brain was registering simple sexual attraction, not love. However, when people were asked how they felt while in the scanner, feelings of strong attachment scored higher than sexual attraction and so, provided the people answered honestly, Bartels might well have succeeded in identifying the love-regions of the brain. If so, then in theory you could scan someone's brain to find out whether or not they genuinely loved you, except of course that they could cheat by thinking of the person they really did love and then the correct areas would still light up.

This research also has a bearing on the importance of oxytocin in love. The regions activated by love happen to be the same areas which have the most receptors for oxytocin. However much oxytocin is present, it can't have an effect unless there are receptors ready to receive it. Rather like a marina, however many boats are waiting out in the bay, they can't drop off the goods unless there's a free berth for them. The brain areas activated by love are like a large harbour with plenty of receptors for oxytocin, underlining its importance in love.

If you wanted to make a love potion it looks as though oxytocin is the way forward. In animal experiments oxytocin is injected directly into the brain but for humans, as well as being dangerous it wouldn't provide the necessary subtlety for a love potion. It could only work if there were a way of getting such a large compound to cross the barrier between the blood and brain. Then of course there would be other practicalities. It wouldn't be like a fairy tale where a person given the oxytocin fell in love with the first person they set eyes on. If you wanted them to fall in love with you then you would need to be the person with whom they were spending the most time. This could be tricky, so perhaps it would be more useful for reinvigorating love. Instead of divorcing, unhappy couples could take the oxytocin and see whether they could fall in love again. An anti-love potion might also be useful. If you were trying to get over a broken relationship you could take something to block the oxytocin and perhaps those feelings of longing would ebb away more quickly.

Taking his research a step further, Bartels asked mothers to lie in the scanner while looking at photos of both their own children and their friends' children. Once again the areas activated by the love for their own children were the regions highest in oxytocin receptors and, what's more, most of the areas were identical to those involved in romantic love. This suggests that attachment between parents and babies is more closely connected to romantic love than we might think.

the strange situation

Chemical similarities might not be all that romantic and parental love share. Adult relationships can also be categorised in the same way as that of toddlers. One person might be clingy and suspicious of their partner while another is more relaxed and independent. The same happens with toddlers and their parents, as Mary Ainsworth demonstrated in 1971 in a classic experiment known as 'the strange situation'.

A mother and her one-year-old baby enter the room and the mother sits and watches her child play with toys on the floor. Three minutes later a stranger arrives, talks to the mother and then to the child. After three minutes the mother slips out of the room without saying goodbye to her child, but she's soon back. Then the stranger leaves. After a few minutes the mother follows and the child is left alone, but soon the stranger returns and so does the mother.

For the baby each stage of these comings and goings is more stressful than the last and by watching the baby's response you can see what sort of bond the baby has with its mother. Ainsworth identified three types, the most common being 'securely-attached'. This encompasses the two thirds of one- to two-year-olds who play happily without clinging to their mother but are distressed when she leaves. When she returns they make straight for her, but soon become calm and continue playing. When the mother is present they are quite happy to play with strangers, but take more comfort from their mothers. This is seen as the ideal and is taken as evidence that the baby has a strong bond with the mother and trusts that she is there for them.

'Anxious-avoidant' is the next type. A toddler of this kind shows no distress when the mother leaves and ignores her return. The baby treats parents and strangers in the same way and is distressed at being alone, but not at the absence of the mother in particular. Because these babies haven't developed a strong bond with their mother, they expect little from them.

In contrast the previous group, the 'anxious-resistants', are loath to explore a room or make contact with a stranger, even in the presence of their mother. When she returns they approach her and may reach out to be picked up, but then struggle to get down again, as though they are punishing her for having left them, something which parents sometimes experience when they collect their toddlers from nursery. The baby wants them back, but then expresses anger at having been abandoned for the day.

Just as babies attempt to secure a bond with parents, adults seek to make a bond with another adult which is so strong that they have no fear of abandonment. Some argue that adult romantic love is simply a type of attachment behaviour that we have evolved in order to stay safe. In terms of survival, the guaranteed protection of one person is very useful.

Adults can also fall into the same categories of attachment as babies, depending on how secure they feel in the relationship. Those who feel secure trust their partners and become more intimate as time goes on. People who are avoidant are supposedly unable to reach such deep intimacy because they don't fully trust their partners. The third group, anxious-resistants, have such a strong fear of what they believe to be an inevitable rejection, that they are unable to form a deep relationship. American psychologists Philip Shaver and Cindy Hazan believe these styles are learnt early in childhood and remain with us in our adult relationships. When they surveyed people to find out which of the three groups best described their adult relationship, they found that they matched the relationships they had had with their parents as children. Of course this won't happen with everyone and some people make deliberate attempts not to model their adult relationships on the way they related to their parents. It was also found that those in the anxious-resistant or anxious-avoidant groups were the most likely to split up after a few years. Once the passion began to wear off they saw less that was valuable in the relationship. The securely-attached people, on the other hand, felt so secure that they did not have those desperate feelings of wanting to be near someone all the time and so passionate love was free to decline.

The popular perception is that although passionate love gradually decreases over time, it is replaced by a companionate love which gets stronger and stronger the longer a couple are together. Rather depressingly, Elaine Hatfield found that this isn't the case. Over the years she's been following a random sample of people in Wisconsin aged between sixteen and ninety and found that although passionate love did indeed decline as people aged, so did companionate love. Sadly many of them are dropping out of her study of long-term love because they are getting divorced. Even amongst those still together most tell her that they are slowly falling out of love. However, there are exceptions: long-term love is possible. Maggie and David, for example, met when they both lived in a civil service hostel in London. She was just eighteen and he was twenty. Both had come down to London for new jobs and with boyfriends and girlfriends back at home, they started out as good friends, but things changed and in 1969 they married. It's not the case that their love has decreased ever since. Today they say they love each other more than ever. They haven't grown apart; they have grown together. Maggie puts part of their success down to the fact that for many years he worked away, only coming home at weekends, which meant they always looked forward to seeing each other.

When retirement loomed two years ago things might have changed. David was to stop work completely and there would be entire days and evenings to spend together. So they planned his retirement carefully. They have separate interests which provide them with enough time apart, as well as plenty to discuss when they are together. Maggie believes that her outside activities with the church and the Brownies are the key. 'You have to be honest about how you want life to be. I spent all my life with a man I've never fully known and he's always surprising me. If you don't refresh your life from outside you don't have opportunities to talk, but that doesn't mean you have to talk all the time. When you're really comfortable together you can have a companionable silence.' It has worked for David too. He plays golf, builds classic cars and busies himself in the shed. Meanwhile he loves Maggie more than ever. 'Fondness and love grow over time. After thirty-five years of marriage we have this incredible sense of comfort together. We become fonder and love each other more all the time.'

Predicting when love will continue to grow like this is not easy. Of all the emotions perhaps this is the most complex. Science can't tell us exactly how it feels, let alone how to find it or make it last, but it can tell us when we're most likely to fall in love, why we might be attracted to one person and not another and why chemicals make us feel the way we do when we're in love.

Although we do know something of what love does to the brain and to the body, most aspects of love still remain a mystery, something which would probably please Senator Proxmire. In fact attraction is still so complex that you might as well date in the dark.

eight

guilt

In 1974 a student called Karl was spending the summer hitchhiking around Europe. Having arrived in Paris in the middle of the night he was walking along a deserted street on the outskirts of the city hoping to find another lift, when a woman came running towards him, screaming in French. Her skirt was torn and both her top and shoes were missing. She must have been attacked. What should he do? He looked around. There must be someone else who could help her. But there was nobody. He wished he wasn't there. He just wanted to be out of the situation. After all, he had his own problems to worry about. He just didn't want to get involved. She came right up to him, crying and talking very fast in French. He pushed her away, ran a hundred yards to the nearest main road and thumbed a lift on a truck. At first he was relieved to have left the scene behind, but ten minutes later as they sped along the peripherique guilt began to overwhelm him. How could he have pushed away someone who was so frightened and helpless? Would she be OK? Why hadn't he at least given her an old shirt from his rucksack or called the police? What must she have thought? Something so appalling had already happened to her and then the person she turned to for help shoved her away. He felt sick with guilt and hated what he'd discovered about himself. 'I didn't want to think I was the kind of person who would just abandon another human in that way, but clearly I am.'

Karl told me this story almost thirty years later, saying that it still haunts him. At first I wondered whether to believe him because it seemed too good to be true, just the sort of story you'd find at the start of a chapter on guilt. Then I felt guilty. Here was someone who had been honest enough to tell a story which showed him in a bad light and I was doubting it. Guilt can catch you at any time.

Guilt is known as a higher cognitive emotion because it involves more complex thought processes than disgust or joy. It's an emotion which lends itself to rumination; people find themselves repeatedly rehearsing the event about which they feel guilty. Meanwhile guilt eats away at them and no amount of distraction can rid them of those gnawing feelings.

Guilt might even have negative effects on the body. At Reading University Professor David Warburton asked volunteers to think of either a happy event in their past or an occasion on which they had felt guilty. He requested that they write down a detailed account of the event (in confidence) before reading it through to themselves. By taking swabs of saliva before and after the task he demonstrated that the immune protein called sIgA rose after people recalled a happy event, but remained the same when they felt guilty, suggesting that guilty feelings suppress the immune system.

Guilt is influenced by the society in which we live to a greater extent than some other emotions. If a particular behaviour is seen as generally acceptable you are unlikely to feel guilty about it. However, within society's moral framework, we also have our own rules; one person might think it's harmless to look at another person's text messages or emails, while others would be appalled. It's an emotion made more complex by the need to assess the cause of an event in addition to the outcome. If your friend breaks their leg you feel sympathy, but if they broke it because you decided to trip them up then you would feel guilty. If you step back and accidentally knock a vase to the floor you would apologise to the owner and feel a certain amount of guilt, but a guilt tempered by the knowledge that it was an accident. According to an idea within psychology known as attribution theory, a person's feelings about an event are dictated by the explanations they give themselves. If you knew you had flung the vase to the floor in temper you would feel far guiltier. People who are prone to guilt carry out an act, blame it on themselves, and then generalise by thinking that they always do the wrong thing in every situation. Making these sorts of attributions can contribute to depression. Another person might look at the context, be annoyed with their clumsiness and try to make up for it, without feeling plagued by guilt or assuming that they must be a bad person in every way.

learning to feel guilty

Two-and-a-half-year-old Willy wandered out of the dining room into the hallway, where he was surprised to see his father, Charles Darwin. At this time they still lived in a house in Bloomsbury in London, though they were soon to move to their long-term family home in the Kent countryside. As his eyes met his father's they flickered, something which didn't go unnoticed by his father. It was the unmistakable look of guilt and the white powder around his mouth betrayed the cause. Although he had specifically been told not to take pounded sugar from the sideboard, he had done so and both his manner and his eyes revealed his guilt. Charles Darwin wrote that Willy's expression could not be due to fear because he had never been punished. Instead he must have been 'struggling with his conscience'. Two weeks later, in the diary of his son's emotional development, Darwin noted that Willy again emerged from the dining room with the unmistakable look of guilt on his face. When questioned Willy insisted there was nothing concealed in his rolled-up pinafore, but later his father found it to be stained with pickle-juice. Once again his guilt was unmasked by his expression.

In theory babies and very small children are unable to feel guilt until they are old enough to make complex judgements involving the idea of personal responsibility. This is something children cannot understand until they are roughly as old as Darwin's child was in the story above. But as with jealousy, Riccardo Draghi-Lorenz from Surrey University believes that psychologists are once again underestimating toddlers' abilities. Draghi-Lorenz thinks they develop a sense of guilt far earlier than the age of two or three, possibly even from birth. His evidence comes from dentition. When breastfed babies grow teeth they sometimes bite down hard on their mother's nipple. This comes as a shock to the mother and she'll tend to cry out in pain, pull away and maybe shout at the baby. For a small baby this might be the first time they have witnessed their mother's anger and the baby will usually cry. Draghi-Lorenz noticed that on subsequent occasions when the baby bites, the baby starts crying without the mother withdrawing. He reasons that the baby, aware of hurting the mother, cries because of feelings of guilt. It is true that babies soon learn not to bite. The problem is that it's hard to guess exactly what a baby might be feeling. Perhaps they cry out of fear, or expectation, having learned to associate biting with a strong reaction from the mother. The mothers to whom I've spoken haven't been convinced that their babies felt guilty about dentition; however they do recall other instances where their babies seemed to know they were doing wrong at a young age. 'From about seven months old Jake's been picking up pieces of fluff from the floor and eating them. He knows he's not supposed to. He even pauses and looks at me, then smiles to himself and does it anyway while watching me warily.' Was Jake feeling guilty or simply testing the boundaries of what he's allowed to do? It is hard to know, but Draghi-Lorenz' work does highlight an important point about emotions. Just because a baby is too young to express an emotion, we cannot take that as evidence that the emotion is never experienced.

Once toddlers can express their feelings of guilt, their understanding of this complex emotion continues to change. In one study children were given two different versions of a story about a bicycle accident. In the first a boy on a bike causes an accident by swerving to miss a child who had run into the road. In the other version the accident is caused by a boy trying to do tricks on his bike in a busy park. Five- and six-year-olds correctly identified which accident was avoidable, but still thought that the two characters would feel equally guilty about the incidents. In another study young children were asked how a boy would feel after he had deliberately attacked another child. Although the children said they knew that attacking another child was wrong they thought that afterwards the boy would feel happy rather than guilty because he had succeeded in doing what he set out to do.

After much work on children's conceptualisation of different emotions, Paul Harris believes that when children first start to understand guilt, they imagine what their parents would think if they knew what they'd done. Then they start to put the onus on themselves and feel that they've not lived up to their own standards if they've done something bad, almost as though they are providing an audience for their behaviour.

The best-known contributor to research in this area was Lawrence Kohlberg. In the 1950s he gave both adults and children moral dilemmas with stories along these lines: a man can't afford £1,000 for the drug that could save his wife's life, but knows that the chemist is overcharging for it. Should he break into the chemist to steal the drug? Kohlberg wasn't interested in whether people said yes or no. What mattered was their reasoning. He proposed that we go through stages of moral development. In the early stages a person might say that the man should steal the drug because if he went there at night he'd be unlikely to get caught. Alternatively he shouldn't break into the chemist in case he gets caught and sent to prison, in which case he couldn't care for his wife. Whatever the decision, this is the most basic kind of moral reasoning which is common until the age of at least nine. At this stage it is the rules prescribed by society which hold sway.

The final stage of reasoning, which not everyone reaches (although Kohlberg of course had), is based on more complex ethical principles. In this case the man is choosing between society's rules and the standards of his own conscience. He might decide that the break-in should go ahead because saving a life is a higher principle than avoiding theft or alternatively that a society functions by an agreement that people abide by certain rules. Therefore he should find another way to obtain the drugs which doesn't infringe the rights of the chemist.

Kohlberg believed that most children would reach the second stage by adolescence and after that every moral decision is reasoned in the same way. Therein lies the problem: Kohlberg doesn't account for discrepancies in higher and lower reasoning. You might decide not to speed in case you are caught and lose your driving licence (lower reasoning), but your reason for not committing murder might be that it is ethically wrong to take someone's life (higher reasoning).

Guilt is also influenced by the presence or absence of other people. In September 2001 I went to Lisbon for a long weekend. One evening I was trying to find the bar where I was meeting somebody, when a man cornered me in a driveway off a busy road. It was dark, but as it was only eight at night there were plenty of people around. As it happens, there were probably too many. I had no idea what this man was going to do to me, but sensed that he was dangerous. I screamed until my throat was sore, but no one came near. He started backing me into the driveway, then fought for my bag, grabbed it and ran. Back on the main road a large group of people had gathered on hearing my screams, but the man sprinted straight through the crowd and down a side street. Nobody tried to stop him and in their place probably neither would I, especially as there were other people around. This is a common reaction. In social psychology it's known as the bystander effect.

There was a famous case in New York in 1964 where a woman called Kitty Genovese was attacked in her apartment. The attacker left the building three times while she was still alive and then returned to kill her. It turned out that thirty-eight neighbours had witnessed the event from their windows, but not one called the police. Each person was aware that lots of other people would be able to see the attack and each assumed that someone else must have phoned the police. The more people who witness an event, the less likely they are to do anything about it. Guilt is also involved. If you witness something alone then it is your sole responsibility and you know that if you fail to help, you will probably, like Karl in Paris, feel very guilty about it. However, if a whole group of you fail to help you need not feel so responsible and the guilt will be less.

the wasted emotion?

Clive and Sarah appeared to be the perfect couple with good jobs, a flat in the city and a country cottage for weekends spent gliding or flying. Unfortunately things weren't so perfect. They both had such busy jobs and social lives that at the age of forty-two Clive realised that they were in fact living separate lives. Although they lived in the same flat, it couldn't be said that they lived their lives together. Eventually Clive decided they should break up and announced to Sarah that he was moving to the cottage permanently. She was shattered by the news. They were doing fine. Life might not be that exciting but they got on well and she was happy.

The guilt of destroying their relationship weighed heavily on Clive, until two months later when a mutual friend told him that Sarah had been having an affair with a married man. It turned out that everyone had known about it, everyone, that is, apart from Clive.

Meanwhile his parents moved to Australia, but keen not to lose touch he would fax them regular letters as well as sending them messages recorded onto cassette. Not knowing about Sarah's affair, when they heard that she was making a trip to Australia they invited her to stay. Clive was furious that they would even consider offering her any hospitality. He told them about the affair, but even then they said that it was too late and the invitation would have to stand. In his anger he told his parents they had to choose between them – it was him or her. He didn't hear from them again.

Eighteen months later he received a phone call from his father's friend to say that his father had been paralysed down one side after a stroke and might not survive. Phoning his mother, he discovered that during the time they had had no contact, in addition to the stroke, his father had also needed two operations for bowel cancer. What's more, his ex-girlfriend had never visited; they had cancelled her stay because they knew how strongly Clive felt about the situation. After this Clive just felt one emotion – guilt.

But was there any point in him feeling guilty? What's done was done. Some would say there is nothing to be gained in torturing yourself with guilt. Sometimes, however, it can be useful. Firstly it can inspire us to make amends or teach us not to repeat the same mistake. Clive flew to Australia to talk to his parents. Luckily his father recovered and since then he's built up a strong relationship with them once more.

Guilt also provides us with information about ourselves and our behaviour. Thinking back to Karl the hitchhiker, he may not have liked what he learned but he did say that through the incident with the girl he had discovered more about himself. Even the thought of feeling guilty might encourage us to do the right thing. You find a nice watch left by a wash-basin at the cinema and you'd like to keep it, but instead you hand it in, knowing that the alternative is to feel guilty whenever you checked the time.

You might expect the people who have done the most wrong to feel the most guilt, but in fact research suggests that the reverse is true. People prone to guilt behave well in the hope of avoiding the feelings. In a classic study back in 1938 ninety-three people were left in separate rooms to solve difficult problems. The answer books were left in the rooms with instructions that some of the books could be used, but not others. Through secret observation it was discovered that almost half the people cheated on the test. Four weeks later they asked people whether they had cheated and half of the cheats owned up. Then they asked everyone whether they would have felt guilty if they had cheated. Only a quarter of the cheats said they would feel guilty, compared with 84% of the others, suggesting that the fear of feeling guilty might have prevented some people from cheating.

Guilt can also be a gesture of appeasement. Expressing

your guilt shows others that you know you've done wrong, which in turn might defuse their anger. As a social species we depend on each other for survival and therefore we need to be able to judge whom we can trust by seeing that others are capable of experiencing guilt. Displays of guilt indicate that a person has a conscience, which will hopefully prevent them from mistreating you.

Emotions don't tend to hold much interest for economists, but the American economist Robert Frank from Cornell University believes that even in business guilt is a key emotion. Economic models tend to ignore emotions while assuming that we make rational, selfish choices, but Robert Frank put it to me like this. Imagine you've lost an envelope addressed to yourself which contains £10,000. If you could pick the person who finds that envelope, who would you like it to be? Unless you know someone who has found this amount of money in the past, you cannot base your choice on previous experience. Instead, you need to choose a person who would give you back the money; someone, who if they didn't, couldn't cope with the guilt. Robert Frank believes that the same process occurs in business. In order to do business with someone you need to know that they would feel guilty if they let you down on a deal.

The classic test of cooperation is the Prisoner's Dilemma which comes from mathematical game theory developed in the 1950s. There are different versions but basically it goes like this. Imagine that you and a friend are arrested on suspicion of robbing a bank. The police put you in separate cells and tell you that if you confess and turn Queen's evidence against your friend you will be set free, while your friend will serve twenty years in prison. If your friend confesses and you stay quiet it will be you spending the twenty years in jail. You're told that if you both confess the two of you will receive a reduced sentence of eight years, but crucially if you both stay silent, you'll be charged with a lesser crime for which you'll both get a year. From an individual's point of view the rational response is to confess, thereby avoiding the worst outcome, the twenty-year sentence. However, if you can be sure that your friend will cooperate you could both stay silent and each get a year, a far better result for you as a pair. It all comes down to whether you can trust your friend. Would they feel guilty if they dropped you in it? If that guilt would be too much for them to handle, you can take the risk. It is, of course, quite a confusing task, so you also have to gamble on you and your friend having worked out the risks correctly.

There are computerised versions of this task and over the years endless experiments have been done, applying the model to everything from vampire bats sharing blood to the decision to push your way towards the emergency exit if a cinema is on fire. This is a situation where you need to guess whether other people will do the right thing. If they are going to fight their way to the exit then some people might be left behind and you don't want to be one of them, so you need to push as well, but if you think everyone else will cooperate you can file out one at a time and maybe you will all survive.

It seems that our brains might even reward us for behaving well towards others. In a recent study James Rilling and his team at Emory University in the United States scanned women's brains while they played the Prisoner's Dilemma. They found that when the women cooperated three areas of the brain were activated, all areas usually associated with reward. So when a person behaved kindly the brain responded with pleasure, but if they played more selfishly it didn't.

the detection of guilt

Working out whether you can trust someone might involve a decision about whether they might be likely to feel guilt in the future. Deciding whether they have already done wrong is rather different. Unlike sadness and fear, guilt has no neat facial expression, but there are clues. One of the most interesting studies using the Prisoner's Dilemma examined one of the ways in which people reveal their guilt – blushing. After screening a group of students at Maastricht University to select the most cooperative and considerate amongst them, Peter De Jong came up with an intriguing task. Two at a time, the students were wired up to a machine which assesses how much they blushed. They were told that the game was a test of their moral strength and that the moral way to play was to cooperate. They were to play for money, holding up a green paddle each time they cooperated and a red one if they didn't (known as defecting). The clever part was that one of the pair had been told in advance they must defect on the fourth round. After running the experiment with many pairs of students they found that whenever a person defected they blushed. This lends weight to the idea that guilt is about appeasement and that blushing is a way of subconsciously signalling to the other person that we know we've done wrong and that we feel guilty. The idea is that the other person then forgives you and conflict is averted.

In the Dutch experiment the students rated each other for trustworthiness and likeability. If blushing plays a part in appeasement, those who blushed should in theory be redeemed in the eyes of their opponent, while those who didn't blush after defection would be disliked and mistrusted. This didn't happen. In fact the opposite happened; the more a person blushed, the less their opponent trusted them. Perhaps they wondered initially whether the person was defecting in order to add variety to the game, but on seeing their blushes concluded that their defection must be down to a deliberate attempt to win money from them.

Toddlers don't blush with embarrassment or guilt until they're about two-and-a-half, suggesting that we need to experience identifiable guilt before we can blush. Adults blush less easily than children. It is not known whether this is because they embarrass less easily or whether it is due to changes in the autonomic nervous system associated with age. Darwin noticed that people often became so flustered when they blushed that the fluency of their speech was reduced. He thought this might be because blood from the brain was diverted to the face. In fact blushing uses only a tiny quantity of blood. Moreover the body is primed to maintain the flow of blood to the brain at all costs, so although you might lose a hand through frostbite the rationing of blood flow to the extremities has protected the brain. So it's more likely that people become flustered not through lack of blood to the brain, but because awareness of their own blushing distracts them.

Not everyone agrees that the purpose of blushing is to indicate self-knowledge of guilt in order to signal appeasement. Darwin insisted blushing is not a form of communication, although his own evidence could be used to dispute this. He noted that in Britain, a relatively cold country, people blush on their face and perhaps their neck and upper chest – the parts that might be visible to others – whereas in hotter countries where people don't wear shirts, they blush all the way down to the waist. In order to reach this conclusion Darwin persuaded several doctors to note down their observations of their female patients blushing while they were examined.

There are two types of blushing – the sudden kind where you realise unexpectedly that your face feels hot, and then there's a more gradual blush. When I'm recording radio interviews I sometimes notice that people who give no other hint of nervousness will blush very gradually down their neck and décolletage. It can take as long as twenty minutes to reach its peak. The blood flow to the face is kept constant by the autonomic nervous system, but when we blush the vessels under our skin have been allowed to dilate and fill with blood, making the face look red and feel hot. Bizarrely these two symptoms don't necessarily happen at the same time. In an intriguing experiment people were videoed singing 'Star Spangled Banner' and then had to sit in a group while the video was replayed – enough to make anyone blush. Meanwhile their skin was monitored for changes in redness and temperature. When the experimenters analysed the timings in detail they found that the face went red before the skin temperature rose. Without a mirror, the only way you know you're blushing is because your face feels hot. This means that other people see you blushing before you are aware that you're doing it.

It's not possible to blush at will, which is why blushing can both provide a useful way of detecting a person's real feelings and can defuse an argument. If the other person sees that you feel suitably guilty they might realise that you didn't intend any offence. It is the very fact that blushing is uncontrollable that allows others to know that your guilt is genuine.

Some people are more prone to blushing than others and the moment they are told they are blushing the problem is exacerbated. Those who blush a lot can find it extremely embarrassing and will even resort to radical treatments such as an operation called an endoscopic transthoracic sympathectomy in which the nerves which dilate the blood vessels in the face are destroyed. It does seem to work quite well; in a recent study of more than 200 people who have had the operation 85% were satisfied with the results. Some patients, however, weren't too happy with the side-effects which can include increased sweating on the torso.

At the other extreme there are people who barely blush at all, remaining cool as a cucumber even if they are guilty. If blushing is supposed to be nature's guilt-detection system it doesn't always come up with the right verdict.

the look of guilt

Darwin cites a number of his correspondents who felt that other parts of the body could reveal a person's guilt. Sir Henry Maine told him that he had observed people in India who gave themselves away when they lied by contorting their toes. As for the face, Darwin believed it was the eyes that were incriminating because they seem restless, as though trying to avoid the other person's gaze. Even when he caught his son taking the sugar, he noted that the look of guilt was manifested in 'eyes unnaturally bright, and an odd, affected manner'.

So what is the best way to lie convincingly and disguise your guilt? Not surprisingly research shows that it is easier to carry off a pre-planned lie than one you come up with on the spur of the moment. Like Darwin, we tend to think that the best way to spot a liar is to watch a person's face for clues like avoidance of eye contact. In fact we are so practised in controlling our faces that we tend to know exactly what expression we are making. Meanwhile our voices can give us away. In studies where people had to watch videos of others either telling the truth or lying, those who were told to pay particular attention to the tone of voice were best at spotting the lies. This suggests that if you want to tell a lie you should do it face-to-face, but if you want to work out whether someone else is lying, phone them up and listen to their voice. But not everyone agrees. Paul Ekman who has spent years researching facial expressions in minute detail says he would prefer to watch someone's face if he wants to know whether they are lying, but only if the stakes are high. He is often interviewed on TV and radio programmes where producers suggest setting up a situation where he has to guess whether someone is telling the truth. He only agrees if the person is willing to gamble a third of their annual salary on the experiment, because it's only when the stakes are high that a person's face gives them away. Simply saying your favourite food is curry when in fact it's prawns, is just too easy because the lie has no consequences.

does the body reveal what the mind conceals?

A man sits at a wooden table with his toes poised in the air and the heel of his shoe on the ground. In his sock there's a sharp nail. When he's asked today's date and for his name and address he presses the front of his foot down hard onto the nail, but disguises his winces of pain. Then he's asked where he was on the evening of 24th February. This time he's careful not to press down. He's trying to fool a lie detector.

It has long been known that guilt can affect the body. Back in 1730 Daniel Defoe wrote, 'Take hold of his wrist and feel his pulse, there you will find his guilt . . . a fluttering heart, an unequal pulse, a sudden palpitation shall evidently confess he is the man, in spite of a bold countenance or a false tongue.'

Lie detectors or polygraphs have in fact had a mixed press and many countries will not accept their findings as evidence in court. They measure changes in pulse rate, breathing and sweating, which are supposedly larger if a person is guilty. Unfortunately being accused of something you didn't do can have a similar effect. It's not hard to spot the pertinent questions in amongst the general ones and this can make anyone nervous. The idea of the nail-in-the-shoe trick is that if you inflict pain on yourself during the bland, control questions then you'll sweat more and your pulse rate will go up, making it harder to find any difference between those readings and those which come up when you

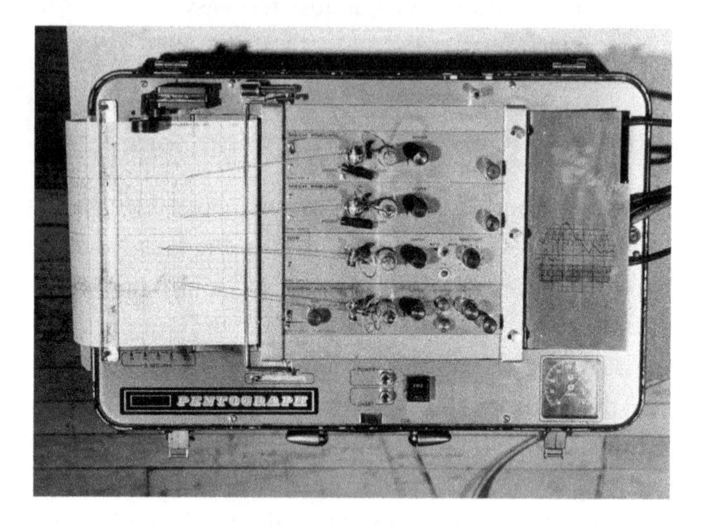

nervously answer questions about the crime. Unfortunately it is the sort of trick that criminals are more likely to have heard about, while someone innocent might fail the test through nerves. During the 1960s David Lykken worked on a secret project for the US Airforce. He trained people to use biofeedback to control their responses to lie detectors. The project was cancelled before he finished but by then he had established that although it was hard, some people could blunt their physiological responses. This gave him the idea of faking the test by doing it the other way around. Instead of dampening down your responses to the pertinent questions, you exaggerate your responses to the bland questions that are asked in order to obtain a baseline. It could even be done without a nail in the shoe. He trained people to do it by biting their tongue, squeezing their bottom, contracting their toes or even doing a complex mental task like counting backwards in sevens from a large number like 436 as fast as possible. Meanwhile people had to take care not to tense the arm or stomach muscles because that could judder the pen on the lie detector and give them away.

Lykken even had the chance to test out his theory in a real prison. A man called Floyd Fay read about his work and wrote to him from prison saying he had been wrongly convicted for murder, partly due to a lie detector test. He was later released, but while he was still in prison he used Lykken's methods to show that nine fellow inmates could easily fake the lie detector test used when prisoners were suspected of breaking the jail's rules. They were all guilty, but came up innocent on the test thanks to Lykken's techniques. Experiments at Utah University confirmed that tests can be faked and, more seriously, that an innocent person has a 50:50 chance of failing a test. Since 1980 Lykken has been testifying as an expert witness in court cases which rely on the results of lie detector tests but finds it hard to convince courts that they are unreliable. People want to believe that there is an easy way of ascertaining guilt.

His alternative approach is the Guilty Knowledge Test. A polygraph is still used but, instead of looking for physiological changes that might indicate lying, the tester looks for changes which might show that the person knows something about the subject. They might ask, 'What is it that you're supposed to have stolen? Repeat each item after me. Twenty pounds, a gun, some whisky, a video recorder, a mobile phone?' The idea is that if the person does know something their autonomic responses will change when they get to that item. It is yet to be used in criminological investigation, but in the laboratory the tests look promising.

hot faces, guilty minds

In the middle of an empty room stands a shop mannequin with a twenty-dollar bill in her hand. A man runs into the room, stabs the mannequin with a knife, grabs the money and runs away. A different man comes in, patches up the mannequin and puts another twenty dollars into her hand. Ten minutes later she's knifed again, as yet another man takes the money. Ten more men follow. Later a group of men, some innocent, some guilty, are questioned. Meanwhile cameras are trained on their faces.

This was all part of a recent experiment conducted at the Mayo Clinic in Minnesota. The volunteers had been instructed to rob the mannequin, but to deny it later. The questioners didn't know which the 'guilty' men were, but they hoped that thermal imaging cameras would help them to find out. Using these cameras, the hottest parts of a face show up orange on the screen. The theory is that when people lie, the skin around the eyes heats up, just as it does when they are startled, possibly as part of the fight or flight mechanism. Using the camera, 80% of guilty people were identified, but worryingly so were 10% of the innocent people. Having said that, the thermal imaging camera still fared better than the polygraph. It is also easier to operate and has the potential for use in public places because people can be assessed without being wired up to a machine. It has been suggested that in the future this technology could be installed at airports to gauge whether people are answering security questions truthfully, although when 10% of innocent people come up guilty this could prove time-consuming.

the guilty brain

It is not the case that each emotion originates from a particular part of the brain. Instead different emotions use different combinations of brain systems. When people with epilepsy have operations under local anaesthesia the surgeon has to stimulate different parts of the brain with an electric current in order to work out where to operate. Because the patient is awake the procedure provides a unique insight into the feelings induced when different areas of the brain are stimulated. A current applied to certain parts of the amygdala can cause feelings of dread, while the application to other parts gives an inexplicable feeling of joy. Touching the temporal lobe can give a deep feeling of guilt. This is curious, since free-floating guilt is supposedly not possible. It must be the most extraordinary experience to feel all these emotions in a row, whilst lying on the operating table. Patients have described these emotions as purer and stronger than they ever feel in real life.

Clues provided by the brains of the guilty are being used in the latest attempts at lie detection. At Sheffield University Dr Sean Spence scanned the brains of people while they either lied or told the truth. He found that truthful responses were given more quickly because in order to lie people needed to use more areas of the brain because they had to both inhibit the lie and create a false answer.

guiltless

Just as some people constantly feel guilty for things that aren't their fault, a few people never feel guilty. This can be a symptom of anti-social personality disorder, in which a person never shows remorse, has no interest in the feelings or rights of other people, is happy to lie and cheat and in very rare cases might commit murder. It's thought that as many as 3% of males and 1% of females might be affected. The biology of this lack of guilt is not fully understood, but one theory makes use of a bizarre link. It's thought that those who behave anti-socially have a low resting heart rate and sweat less than other people. The same physical attribute is often found in bomb disposal experts and just as this enables them to keep calm, so it might influence a person intent on committing crime.

Adrian Raine, a psychologist who has worked in high security prisons, wanted to investigate this idea so he took physiological measurements from English schoolboys at the age of fifteen and then kept track of them for the next nine years. Extraordinarily he found the boys with low physiological arousal were the boys who grew up to commit crimes. Perhaps this physical calmness means they miss out on the body's signals that something is wrong. This in turn stops them pausing to consider the consequences of an action or to feel any guilt. To get any sensation of excitement they might need to do something drastic – shoplifting, for example. At the moment the links are tenuous. Reduced bodily sensations might explain a lack of fear during violence, but cannot explain the lack of guilt or remorse afterwards. What's more, not everyone with a low heart rate turns to crime. Maybe the fact that their heart rate remains slow during a crime is a symptom of their lack of concern, rather than the cause. If you are not worried about the consequences of a crime you won't feel anxious.

This brings us back to the century-long debate between the theories of William James and Walter Cannon. The question was: which comes first, the emotional experience or the body's reaction? Common sense would suggest that we feel afraid and then notice our hearts beating faster, but William James said it was the other way around, that we experience emotions *because* we are aware of changes in the body. At first sight this might seem daft, but there are situations where you might have noticed it happening in this order. You're at a wedding waiting for your turn to give a speech. You've practised the speech at home and you know everybody well so you're not too worried, until moments before you have to stand up. Suddenly you can feel your heart beating hard inside your chest and you feel your palms beginning to sweat. Then you feel nervous and wish you didn't have to do the speech.

The physiologist Walter Cannon insisted it happened the

other way around. He claimed it was impossible to induce an emotion in a person simply by altering their physiology in some way. More recently he's been proved wrong. I've already described the facial feedback experiments where people were made to move their facial muscles into a smile and felt happier as a result. Another intriguing experiment was conducted by the social psychologist Stuart Valins. He rigged people up to a machine that let them hear their own heartbeat, or so they thought. In fact he played them an artificial heartbeat, the speed of which he could control. He showed the men a series of pictures of half-naked women accompanied by a heartbeat either slower or faster than their own. Afterwards those who believed their heart to have been beating fast rated the women as more attractive than those with the slower heartbeat. Their false perceptions of their own bodies' reactions were influencing their feelings.

One of the problems with the theory that your body informs your feelings is that the same physical movements don't always conjure up the same emotions. If you are running fast because someone's chasing you then fear is the most likely emotion, but if you are running fast towards the finishing line of a race aware that your competitors are behind, you might feel elated. The way we view a situation clearly has an influence on our feelings. In the experiment with the false heartbeat machine, a person's genuine physiological response was successfully masked, suggesting it's our perception of our symptoms that matters rather than the symptoms *per se*. Those perceptions are inevitably mediated by our thoughts and beliefs. It is also hard to see how the body could induce more complex emotions. What could the body do to make you feel guilty?

This has been a huge topic in emotions research, with endless experiments. The answer is probably that it works both ways – emotions give rise to physical sensations and physical sensations inform emotions. Perhaps the physical symptoms force you to focus on the emotion you're experiencing. This intertwining of the physical and the emotional can even be circular. If you think back to the research on blushing, they found that people blushed before they felt hot and became aware of it. Once you are aware that you are blushing, you feel worse and even more embarrassed which in turn makes you blush more and then feel even worse.

survivor guilt

Ben Gillow was a seventeen-year-old working as a courier when he was given the job of flying from Bristol to Basel on 19th April 1973. As they came in to land in Switzerland the weather was closing in and it began to snow. Unable to see the landing beacons, the pilot overshot the runway, banked steeply and looped back around to the airport. Time for a second attempt. There was the first beacon. Then the second and then nothing. Once again no third beacon appeared and he had overshot the runway. In the cabin the atmosphere was particularly jolly because most of the people on the flight were travelling together in large groups. Nobody seemed too concerned about the multiple landing attempts. They just laughed it off. It was bound to work the next time – third time lucky. But it wasn't to be. Visibility was so poor that the beacon the pilot assumed to be the first was in fact the second. He was landing too late every time. The plane touched down again, then reared into the sky but this time, as the plane climbed back up, one wing clipped the edge of the mountain. The lights went out and Ben was thrown forwards as the plane lurched. Within seconds the plane hit some fir trees which flipped it over, and it crashed into the snow with its tail wedged into the ground.

When Ben came to he was hanging upside down in his seatbelt. After freeing himself from the seat he found an emergency axe and broke down the door, calling to the other survivors to come to the back of the plane. He stood in the snow, helping to catch people as they jumped out. They sat in a group some distance from the plane and waited. Two hours later they were still waiting. Surely the airport must have been aware of the plane's difficulties. They must be out looking for them. As the youngest, fittest man in the group Ben set about building a fire and then returned to the plane to look for alcohol and matches. While he was there he painstakingly clambered along the upturned plane to check every row of dead bodies hanging upside down just in case anyone was still alive. Some had been decapitated, but when he continued to look he found a mother and son who were still breathing. He managed to get them out of the plane and carried them back to the group of survivors. Many of them were so shocked that they simply sat in the snow, but Ben was determined to get help. Eventually he saw a boy walking in the distance with his dog and despite his three broken ribs he ran to catch up with the boy, walked half a mile with him through thigh-high snow to the nearest house and phoned for help.

Years later the crash still haunts him, but not because of what he saw. Today he cannot even picture the dead bodies in the plane, as though his mind is protecting him from the horrific scene he witnessed. What haunts him is the fact that he had survived when 108 people had not. Why did he deserve to be saved? 'I was seventeen with no responsibilities. Why should I survive when a woman with five kids doesn't?' Although he was in no way to blame for the crash he felt guilty for his survival. In fact he had been the hero, breaking down the door of the plane, helping the passengers to escape, rescuing the mother and son, and raising the alarm, but his chief emotion was survivor guilt.

The term 'survivor guilt' was coined in 1964 to describe the particular sort of depression found in some survivors of Nazi concentration camps. It can be present after any incident where not everyone survives. In the Piper Alpha disaster in 1988 160 men died in the worst oil rig disaster there has ever been. Many of the 59 survivors slid down pipes and jumped into the sea to escape the flames. Just over half saw somebody else either die or receive serious injuries and most thought they would die themselves. More than ten years after the accident the psychiatrist Alistair Hull followed up the survivors. A third still felt guilty that they had survived when so many others had not.

Survivor guilt involves the suffering of others who were both in a comparable position and no more deserving of bad luck than you. At first sight this kind of guilt appears to be useless – a wasted emotion. Nobody can change what has happened and those who died would not have wanted their friends or colleagues to spend years consumed by guilt. Strangely, survivor guilt might in fact help people to deal with what has happened, by suggesting that they do have some control over events. After an accident we often feel very helpless because it highlights just how unfair life can be. Anything which might help us to regain a feeling of control over events could help us to feel that most useful of emotions, hope. As we'll see in the final chapter, we seem to need optimism in order to function. When an accident happens out of the blue it blows a hole in our idea that everything will be all right. If we believe we do have some control over events, we might feel a bit better. The downside is that imagining you had some power over events inevitably leads you to feel guilty for surviving when others didn't.

Richard Blacker, a doctor from Boston, noticed that patients sometimes become depressed after heart surgery. Many people underestimate their chances of surviving cardiac surgery, but their surprise at survival is often clouded by a bout of depression that begins three or four days after surgery. Blacker found that patients often had siblings or friends who had died through illness at a younger age than them and it was survivor guilt that seemed to be the key to their depression. It is also common in the recipients of transplants, who are all too aware that their survival depended on someone else's death.

Forty years after leaving Auschwitz the chemist-turned-writer, Primo Levi, was still suffering from survivor guilt:

> I might be alive in the place of another, at the expense of another; I might have usurped, that is, in fact killed. The 'saved' of the Lager were not the best, those predestined to do good; the bearers of a message. What I had seen and lived through proved the exact contrary. Preferably the worst survived, the selfish, the violent, the insensitive, the collaborators of the 'gray zone', the spies. It was not a certain rule (there were none, nor are there certain rules in human matters), but it was, nevertheless, a rule. I felt innocent, yes, but enrolled among the saved and therefore in permanent search of a justification in my own eyes and those of others. The worst survived – that is, the fittest; the best all died.

In 1987 Primo Levi was found dead at the bottom of the stairwell at his home in Turin. He left no note, but was deeply depressed and is suspected to have killed himself. That morning he had phoned the Chief Rabbi of Rome and told him 'I don't know how to go on. I can't stand this life any longer. My mother has cancer, and each time I look at her face I remember the faces of the men lying dead on the planks of the bunks in Auschwitz.' Despite the international recognition he received for telling the story of the people who died in Auschwitz, he insisted that his testimony was only that of a survivor and that the true witnesses, the people who died, had never had the chance to tell their story. His feelings of guilt haunted him for the rest of his life and maybe contributed to his death. In this case it is hard to see how survivor guilt was useful.

Even with therapy it can be hard for people to let go of

their survivor guilt. Tom Williams treats Vietnam veterans at the Post-Trauma Treatment Center in Colorado. He feels that survivor guilt sometimes helps people to avoid thinking about the horrific facts of an event. Instead they spend time going over the scene imagining what they could have done differently. Williams doesn't confront veterans directly on the subject of guilt. Instead he encourages people to accept that their sadness is a reasonable way to feel after the experiences that they've had, whilst promoting the idea that the deaths of others were not their fault. He asks them to consider how the decisions they had to make in Vietnam were made with little information under extreme time pressure. He has found that many soldiers were so young at the time that they still had adolescent views of right and wrong. He tries to help them to see that the situation wasn't black and white. Some soldiers feel guilty that the life they've led since Vietnam has been chaotic and that they have somehow failed in their duty to justify their survival by leading a good life. Williams asks each veteran to recall the ways in which they tried to help their comrades during the war. One soldier found marks on the ground and realised they were made by his friend's boots when he was dragged away. For three days he followed the tracks in the hope of finding his friend and when he did find him it was too late. He had been killed just before he caught up them. Understandably he was haunted by the idea that if only he had moved faster, he could have saved him. Williams encouraged him to take pride in the huge efforts he had made to save him, reminding him that his friend would have been pleased that someone had tried so hard to save him and would have been grateful to his friend for recovering his body for his family.

Cases like this one and Primo Levi's are extreme, but survivor guilt can take milder forms. For example, sometimes people feel guilty after passing an exam while their friends fail. Some researchers have suggested that mild survivor guilt can be useful to society. The idea is that within society we constantly make comparisons between ourselves and others, with the aim of either improving or maintaining our status. At the same time, everyone knows that if they rise too high they might become disliked. By feeling guilty for achieving success it is almost as though a person is trying to put themselves back in their place. This stops the most successful people from showing off too much, allowing everyone to coexist peacefully.

While severe survivor guilt remains a mystery, basic guilt does seem to have its uses, whether it's to help us learn from our mistakes, to prevent us from behaving badly or simply to promote cooperation by allowing us to trust each other. It is an emotion which can stay with people for an extraordinarily long time. Thirty years after refusing help to the woman in Paris, Karl still feels guilty, but in an attempt at redemption, he now goes out of his way to be helpful, picking up hitchhikers and dropping them wherever they want to go. He says he can't help the woman he abandoned, but he can make it up to the world in general.

nine

hope

It was a wet May afternoon when I arrived in the French Pyrenean town to be faced with street after street of gift shops selling plastic tat. It was soon clear that each visitor follows the same routine. First they go into a gift shop to buy a plastic bottle. The size selected is in proportion to their age. Small children choose tiny hotel-shampoo-sized bottles for half a Euro, while old ladies struggle with five-litre billy-cans. Then they walk, wafted along by the crowd, towards a Disney-like palace fronted by two grand swirling staircases, but the palace is not what they are looking for. They walk on past, all knowing exactly where they want to go. They head for a line of metal cages piled up with white candles. Two Euros in the slot and you take your candle. Again there's a choice of sizes. Then they queue up quietly in the rain to touch a piece of rock. Already they know that the daily parade is off due to the wet weather. Standing in the pouring rain, aware that the main attraction is cancelled, would be enough to make you give up hope but in fact hope is very much in the air and it would take more than rain to spoil it. This is Lourdes and everyone is hopeful.

Those who are ill or have disabilities sit in giant blue carts with pram-like hoods, pushed by nurses dressed in cloaks, A-line skirts, neat purse-belts and white, starched hats. It looks like a scene from the First World War. Some people are so ill that they are wheeled along on narrow, flat trolleys, with drips hanging from stands above the beds. It should be depressing but nobody looks sad. It's raining

harder but still everyone queues contentedly. When they finally get their turn inside the cave where St Bernadette had a vision of the Virgin Mary, they touch the rock keenly. The people using wheelchairs reach behind them, brushing the rock with their hands to ensure they touch the cave for as long as possible. As people emerge from the cave I can see their faces; they are clearly thrilled.

Afterwards people head down a slope into what looks like

an underground car park. It has a concrete roof held up by rows of buttresses, as though we're gathering under the belly of a fifty-legged beetle. The people in bright blue wheelchairs line up at the front with their pram-hoods lowered. Every day for six months of the year thousands of people attend a service held in multiple languages. The priests' images are bounced up onto big screens as though at a rock concert, so that wherever you are standing you can see what's going on. At certain points during the service the priests speak and the congregation reply as one. A select few are even pulled up onto the stage to read passages into microphones.

Not everyone who comes to Lourdes expects a miracle, but even as a non-Catholic I could feel the sense of belonging that people share here. Despite so many seriously ill people gathering in one place the one emotion absent was despair. Instead there was a warm feeling of hope.

Hope is a tricky emotion to pin down. Some would argue that in scientific terms it's not an emotion at all. There isn't a distinct facial expression for hope; a look of hope is easily confused with signs of interest or religious feeling. It's a quiet emotion which often gets forgotten, but it is a feeling we all recognise. Hope changes the way we think. People who are hopeful evaluate themselves more positively and generously than they judge others. At first sight hope might not appear to have physical manifestations. It doesn't raise your heart rate in the way that fear does. Unlike guilt, it doesn't make you blush, but through changing the way we think hope can have radical effects on the body and, as I'll cover later in the chapter, it can even postpone death.

* * *

Hope, like fear, involves the anticipation of a possible outcome. Seneca wrote, 'You will cease to fear if you cease to hope. Both belong to a mind that is in suspense.' Both fear and hope deal with uncertainty, but whether that uncertainty breeds fear or hope depends on the way you view a situation. You cannot hope for something without also fearing it might not happen. So with any uncertainty you can choose whether to fear a bad outcome or hope for a good one. It's the old idea of deciding whether a glass is half full or half empty.

the importance of hope

Hope forms a basis for much of life. Without the hope that a person will treat us well we would never start a relationship. We hope and trust that the other person will be nice to us. Couples go to great lengths to have children, with no idea what those children will be like or how they will find life as a parent – a real act of hope and an act of hope in a larger sense as well; that there will be a future world that is worthy of their children. Hope motivates us to plan for the future and to strive for the things we think will make us happy.

James Averill, one of the few emotions researchers to focus on hope, asked people to think back over the last year and to pick an episode which for them represented hope. It could involve any situation, provided it had started and ended during the year. Just over 40% described an event relating to an achievement such as getting good grades, finding a good job or winning at sport. A quarter had hoped to find a partner, to marry their current partner or to get on well with friends and family and 8.7% had hoped for something to happen to someone else – recovery from illness, for example. The rest had hoped for things ranging from a more outgoing personality to a new car. These results demonstrate the variety of occasions where hope is deployed. When Averill asked people how confident they were that their wish would come true, regardless of when he asked them, the average estimate was that it was 58% likely to happen. Perhaps a definition of hope is that something needs to be marginally more likely to happen than not.

People who consider themselves to be lucky also tend to be the most hopeful about the future. Not surprising, you might think. If they've had a lifetime of things going right for them, it's no wonder that they will anticipate more good times in the future. However, this could be a self-fulfilling prophecy. Richard Wiseman started a luck school at the University of Hertfordshire where he trained people to become luckier. Through various experiments he had found that people who considered themselves to be lucky behaved differently from supposedly unlucky people. In one study participants were asked to look through a newspaper and count the number of photographs. The lucky people took only seconds while the unlucky people tended to take about two minutes because they failed to notice a half-page advert on page two which read 'Stop counting – there are forty-three photographs in this newspaper!' Wiseman tried it with a different message. This time it said, 'Stop counting, tell the experimenter you have seen this and win $250!' Again it was the self-professed lucky people who spotted the advert.

Wiseman believes that lucky people make their own luck by behaving in particular ways. They are better at noticing opportunities, they make lucky decisions by listening to their intuition and they expect things to work out. When life goes wrong, they see that as an exception. At his luck school he taught 'unlucky' people to behave differently. Afterwards 80% said they felt luckier and some even found new partners and new jobs. This suggests that being hopeful isn't simply a way of looking at the future. It can actually change it.

One method people use to increase their feelings of hope is superstition. If there's something you can do that will help to make things go right, whether it's avoiding or seeking out black cats, then you can feel more hopeful. Back in 1948 the famous psychologist B. F. Skinner discovered that pigeons are so 'superstitious' that even when food is delivered to their cage every fifteen seconds, they will repeat whichever movement they were making when the food last appeared. By the end of his experiments he had one pigeon which constantly walked in anti-clockwise circles and another which repeatedly thrust its head into one corner of the cage. Humans also make connections between their actions and the events that follow them. Although these connections do not really exist, they do give us the semblance of control over what happens to us. Interestingly, far more people act on positive superstitions than negative. For example of lot of people consider it daft to avoid walking under ladders, but still touch wood or cross their fingers when they're hoping something will happen.

From the age of about eight our level of hope becomes

fairly constant but it is hard to know when we first hope. Does a hungry baby hope that someone will respond as soon as it cries? Wittgenstein said that you can't observe a child and wait for it to manifest hope. Instead he believed that a child's life gradually changes until there is a place in it for hope. You could argue that for an infant to feel hope they need to know what their goals are, what possibilities there are and that they have some agency in the world, that they can make things happen.

These constituents of hope fit in with the theory developed by Rick Snyder, the world authority on hope. He runs a university department in Kansas, where he and his colleagues and students come up with endless new aspects of hope to investigate. In his theory, to feel hope you need to have a goal; you have to hope for something in particular, although it need not be an object. You could hope for something as broad as feeling happier in the future. Coupled with this goal you need to be able to see a possible route to achieving your desire and believe you have the potential to do so. Once you have all three of these elements you can experience hope. New hope is energising, which is why people feel so good when they have a new plan. People starting diets have been found to be particularly happy because that sense of hope that things are going to change is extremely powerful. Hope can even spread to an entire community, brought on by a change of government, for example.

People who score high on hope also score high on optimism, and optimism and hope do share many similarities. Individuals vary considerably in their levels of optimism, regardless of how many dreadful things might have befallen them in life. Chris Peterson from the University of Michigan has devoted his career to studying optimism. He distinguishes between little optimism and big optimism, saying that people might be optimistic about the small things in life, like whether the train will be on time, while they're pessimistic about the bigger picture; 'the world is becoming a crueller place'. This distinction does seem to be useful. I, for one, tend to be optimistic about the long term, but when it comes to a short bus or tube journey, I think it will probably go wrong. Based on previous experience I think I have particularly bad luck with transport, despite the lack of any logical explanation for this. The people running the tube don't see me coming down the escalators and then change the waiting time from one minute to thirteen, but some friends of mine are unaware the digital displays go into double figures because they are always so lucky. If I travel with them it all goes smoothly, leaving me with little hope when I travel alone. Clearly I'm not good on little optimism.

There is a long list of ways in which hopeful people benefit. They are more likely to succeed, whether at work, in sport, in academia or in politics. They are also happier and better both at persevering with a task and solving problems. Hope is useful in every area of life where effort brings benefits. Part of the reason why hopeful people succeed is that as well as setting themselves higher goals, they set more goals at a time. This also makes them happier because it buffers them against disappointment; if they fail in one area they remain cheerful because they still have the others. When disappointments do come, both optimistic children and adults are less likely to become depressed.

In one of Snyder's studies, students' feelings of hope when they started at college were a better predictor of their final college results six years later than their entrance exam marks. Levels of hope even foretold the students who were later to drop out. Snyder believes that if their levels of hope could have been enhanced it could have changed the course of some of the students' lives, because hopeful people are more motivated and work harder.

Optimistic people not only achieve more but they have better health and are more popular. As with joy, where happy people smile more and smiling makes them feel even happier, there's a cycle with hope. If you are more optimistic you are more likely to succeed and that success in turn will make you even more hopeful the next time, based on your past experiences.

We are also influenced by the level of hope surrounding us. If everyone else is optimistic about a project we are likely to pick up on that optimism. Likewise hopelessness spreads fast, hence the introduction of sedition laws during wartime where simply suggesting that your country might lose a war can be a criminal offence. This is one of the reasons why it is so hard to raise a group's morale once it has dropped. Sometimes hope is all that people have. When Mark Henderson was released three days before Christmas in 2003, after 102 days spent as a hostage in the Colombian jungle, he said that although his emotions see-sawed from day to day what kept him going was hope.

Even negative events can increase hope. Just nine days after the attack on the World Trade Center on 11th September 2001, Barbara Fredrickson conducted a study using college students at the University of Michigan. A few months earlier the same students had been assessed on various characteristics including resilience and levels of optimism. Surprisingly they found some students felt more optimistic after the terrorist attack than before and these were the students found to be the most resilient in the first study. In national polls only 21% of the population felt optimistic after 11th September compared with 68% eleven years earlier, but somehow the most resilient people not only coped, but came out feeling better than before. In fact, when tragedies occur many people cope better than they would expect.

enduring hope

Rachel Hedley said goodbye to her husband Charlie one January morning in 2002 after he had dropped their two children at school. He set off for his daily bike ride to work as a road safety specialist in central London, but he never arrived at work that day. On his way in he was involved in a collision with an articulated lorry and he died. He was a very keen, experienced cyclist and his death at the age of only thirty-nine came as an enormous shock to his family. This story might appear to fit better into the chapter on sadness, but it is included here because Rachel still considers herself to be a lucky person despite losing her husband. Somehow she still manages to look on the bright side, just as she always has done. She tells herself she was lucky to have met him at all, lucky to have had eighteen years with him rather than eighteen months, and lucky to have had such great children with him. In her bleaker moments, she finds that thinking about good things that happened in the past brings her hope.

Her hopefulness has led her to plan events which make a difference to her own future. Charlie had intended to spend his fortieth birthday cycling up the Col du Tourmalet, one of the high Pyrenean mountain passes famously used as a climb in the Tour de France. Rachel decided that she would do the ride in his place. She recruited sponsors and lots of her friends decided to join the ride. By the time they reached the café full of cycling memorabilia at the top of the mountain, rather than feeling grief-stricken, they felt fantastic. She said that on reaching the top of a high mountain she couldn't help but feel on top of the world. Charlie's mother arranged a big party in the evening and despite the fact that they were celebrating the fortieth birthday party of someone who had not lived to see it, Rachel had turned a tragic day into an event she remembers with fondness. Without hope she would never have contemplated the trip.

Rachel's story illustrates the fact that hopeful people have not necessarily had fewer bad things happen to them. It's the way they look at things which makes a difference. Essentially when something good happens they take credit it for it and when something bad happens they blame outside circumstances. These attributions needn't be public. If, for example, a person is responsible for a member of staff who ruins a project, the person might need to take the blame publicly, but what affects their sense of optimism is whether they actually believe it was their fault. If you are late for an appointment do you blame yourself for never leaving enough time to reach places, or do you simply think you were unlucky because there was an unusually large amount of traffic on the roads? Conversely, if you arrive at just the right time do you think how lucky you were with the traffic or inwardly congratulate yourself for your good time-keeping? If you want to be optimistic you need to do the latter; take inward credit for anything good that happens and no responsibility for anything bad. This is at the heart of the idea of learned optimism, a theory developed by Martin Seligman in the United States. He found that hopeful people believe that good events reflect permanent qualities of their personality and are likely to happen again, while bad events are seen as one-off bits of bad luck. This is known as a person's explanatory style and, as I mentioned, it can contribute to a person's tendency to feel guilty. Every judgement a person makes has three important components – is it external or internal (caused by you or by outside factors), does this just apply to this circumstance or is it global (for example, you dropped the glass because you were distracted for a moment – or did you drop it because everything you do is stupid and clumsy?) and finally, was it stable or unstable (a one-off event or something that will happen again)?

These ideas came out of a theory called learned helplessness. In a particularly unpleasant rat experiment it was found that a complete lack of hope (or helplessness) could affect an animal's ability to survive. The study was conducted in the 1950s by an American psychologist called Carl Richter. He put a rat into a three-foot-high glass jar of water, specially designed with a jet of water which poured down onto the rat's head to prevent it from floating. The only way for the rat to avoid drowning was to swim continuously and it was found that they could do this for sixty to eighty tortuous hours. Then Richter took another set of rats, but this time he induced feelings of helplessness by grasping each rat in a chain-mail glove until it stopped struggling. When he put these rats into the jar they would only swim for between three and five hours before dropping to the bottom and drowning. He believed they died from a lack of hope. If he rescued a rat part way through the experiment and then returned it to the water, hope was regained and the poor rat would swim for sixty to eighty hours like the others.

hope and health: a gift from arden house

Arden House Nursing Home in Connecticut had a good reputation. It was large and modern, spread over four floors. Two of the floors were picked at random and it was announced to the residents of these floors that there were some new plans for making their lives at the home as comfortable and pleasant as possible. The residents on the second floor were each given a present from Arden House, a plant which they could keep in their room. A member of staff handed out the plants, telling the residents that the nurses would water them regularly. On Thursday and Friday evenings there was to be a film-showing and later the residents would be told who was scheduled to see the film on which night. The forty-four residents were all looking forward to the changes.

On the fourth floor the plans were the same but for a

couple of subtle changes. Residents were given the choice of whether to see the film on Thursday, Friday or if they preferred, not at all. The box of plants was passed around for everyone to select their own plant as the present from Arden House. They were told the plants were theirs to tend and also that they could decide on the arrangement of furniture in their room and that they should feel free to tell the staff if they had any complaints or could think of anything they'd like done differently. Ellen Langer and Judith Rodin, who led the research, gave the residents questionnaires to assess how happy they were at the home.

Both groups were treated kindly by the staff, but after eighteen months there was a distinct difference between the two floors; although the two groups were of similar health at the start of the experiment, twice as many of the first group died during the study. The only difference was the degree of control they had over their lives. The people who looked after their own plants led an equally restrictive life, but were able to make small decisions for themselves. Meanwhile those who had everything done for them became more dependent and felt more helpless and hopeless. As with the rats in the water experiment, hopelessness hastened death. Even a minimal feeling of control over your situation can instil hope.

Hope can also help us to cope with stress. During the Second World War psychologists realised that bomber pilots coped better with the dangers they faced if they knew that they had a limited number of missions. Otherwise they would have to continue until they were either dead, captured or injured, whilst watching more and more of their colleagues die. It was decided that they should fly thirty operations, have a break for training and then fly another twenty. In theory this gave them a 50:50 chance of survival, but the knowledge that the missions were limited successfully reduced the number of mental breakdowns. The only problem was when a pilot only had a few more missions left to fly the strain would begin to show and the final flight was the most stressful of all. According to the psychiatric historian Ben Shephard superstitions were quick to develop. Men would carry lucky mascots, while any woman unfortunate enough to have been out with several pilots who had since died became known as the 'chop-girl', who was then avoided by everyone else. It was hope that kept the men going; they simply hoped that they would not be the ones to die.

the nun study

The Mother Superior at the School Sisters of Notre Dame in Milwaukee sat down at her desk on 22nd September 1930 to draft a letter to each sister who was preparing for a life in the religious order. First as a postulant and then a novice, each was about to vow to dedicate her life to the mission of Jesus. Each would swap her white veil for a black one and at the service she would wear a crown of thorns before lying face down on the floor in front of the altar. The whole group would be covered in a black shroud to represent the death of their old selves. Then they would take their vows of poverty, chastity and obedience. In her letter the Mother Superior asked each sister to write a short sketch of her life to date in which she was to include her place of birth, details of her parents, 'interesting and edifying events of childhood', schools attended, influences that led her to the convent, religious life and 'outstanding events'. It should only run to a page and it would help the Mother Superior to choose the appropriate career path for each nun, as well as providing the order with information for use in obituaries much later on.

Little did these sisters know the extent to which the page they wrote about themselves would be scrutinised seventy years later. It might even provide some of the secrets to long life. In the 1990s a world expert on Alzheimer's disease, David Snowdon, approached the order and was given permission to follow the health of 678 nuns, all born before 1917. The nuns even agreed to donate their brains to his research after their deaths. For an epidemiologist the nuns provided the perfect research sample because the restrictive nature of their lives meant they had a huge amount in common; they didn't smoke, have sex or children and they had similar diets, jobs, healthcare and income. Most of the research in the nun study has looked at the development of Alzheimer's, but Snowdon also wanted to know what factors were causing the differing lifespans of the nuns. Their lifestyles were so similar that something else had to be making the difference to their longevity. When the order showed Snowdon the material in their archive he was amazed to read the set of autobiographical statements written so many years earlier. Could these letters from the 1930s hold the key to long life? Each statement was analysed to see how often in their descriptions of themselves each nun used positive emotions like hope and happiness, neutral emotions like surprise and negative emotions such as anger. When the nuns wrote their statements their average age was twenty-two, but more than sixty years later it was found that the nuns who had included the greatest numbers of references to positive emotions in their statements were the same nuns who had lived the longest.

This result is extraordinary, especially when you bear in mind that when the nuns wrote these words they had not been asked to discuss their emotions. They were about to take their vows so you might expect them all to have been looking forward with hope to the future and to want to present their best side to the Mother Superior. The quarter of the nuns who used positive emotional words the most often lived for an average of 9.6 years longer than the bottom group. This is even longer than the average difference in life expectancy between smokers and non-smokers.

This is not the only evidence that hopeful people live longer. In the United States a group of people were tracked across thirty years and the optimists were both prone to less illness and lived longer than the pessimists. Both groups were equally healthy at the start of the study, but more illness emerged later in life in the pessimists.

Hope also appears to play a role in the course of disease. In a study of women with breast cancer Dr Steven Greer found that those who responded with a 'fighting spirit' had a better prognosis while those who were filled with hopelessness fared worse, even when the severity of their cancer at the beginning of the study was taken into account. However, patients who don't feel able to approach their disease with a fighting spirit need not feel guilty. There was another attitude in this study which had the same success rate – denial. This did not mean that patients necessarily denied that they had cancer altogether. Instead they coped with the idea of the disease by minimising its seriousness, for example, convincing themselves that an operation had been purely precautionary or saying to themselves, 'There might have been a few cancer cells, but there's nothing serious to worry about.' This approach was just as effective in fighting the disease.

After his discoveries Dr Greer developed a type of therapy called Adjuvant Psychological Therapy, with the specific aim of helping patients to develop a fighting spirit. Denial tends not to be encouraged, which is interesting considering the results of the study. Perhaps doctors would find it hard to deal with patients' denial. Any counselling to encourage a fighting spirit needs to be conducted with great care, for fear of erroneously implying that a person can think their way out of a disease as serious as cancer. If their disease worsens they might blame themselves. In Greer's study, even with a fighting spirit only four out of ten women were still alive fifteen years later, although intriguingly another four died of causes other than breast cancer. If denial is as effective as a fighting spirit, perhaps it is the presence of hopelessness which accelerates the disease process rather than hope itself slowing it down.

Hopelessness can certainly lead to death. Everyone knows stories of elderly people who die within weeks of their spouse and there is a recorded increase in the death rate amongst men over the six months after they have been widowed. After this time the death rate seems to go back to normal. If hopelessness can lead to death it's no wonder that lack of hope has long terrified us. From John Bunyan's 'Slough of Despond' in *A Pilgrim's Progress* to J. K. Rowling's Azkaban prison where jailers sucked the hope out of prisoners, leaving despair in its place, it has long been realised that hope is essential.

While an absence of hope can shorten life, its presence can prolong it. In the week before the turn of the millennium fewer people died than usual, while in the week after the new year there were more deaths than usual, rebalancing the numbers. Some people were literally hanging on to see in the new millennium.

It is thought that the reason hope or the lack of hope can affect health is because of its impact on the functioning of the immune system. Optimists even recover from surgery faster than pessimists. Snyder found hopeful people engaged in more preventative health behaviours such as taking exercise. This is interesting because you might suppose that hopeful people would assume illness was unlikely to befall them and would therefore be less likely to bother taking preventative action. Amongst Snyder's extraordinary range of studies of hope he has even found that higher hope can help people to deal with burns injuries, while others have found that it makes it easier for people to cope with spinal cord injuries, severe arthritis and even blindness.

Hope can also affect your pain tolerance. Shlomo Breznitz asked people to plunge their arm into a bucket of iced water, a common test for pain tolerance because the exercise proves enormously painful surprisingly quickly. One group of people was told the test would last for four minutes and the other was simply told to keep their arm in the water for as long as they could. In both cases the test was stopped after four minutes, but only 30% of those who were unaware of the length of test could stand the whole four minutes, while 60% of those who knew they had only to last for four minutes, stuck out the whole test. The same principle is exploited in all sorts of situations. In an exercise class, the teacher will wait until the exact point where you think you cannot possibly do any more sit-ups and then say 'only eight more to go!' and, knowing you've almost finished, you are able to push yourself that bit further. Had you been on your own you would have given up.

instilling hope

All these studies suggest that hope is a good thing, helping us to feel happier, achieve success and even live longer. In theory, if you can see the good side in something bad you can turn an unfortunate situation to your advantage. For example, you view the loss of your job as an opportunity rather than a disaster. Of course this is far easier said than done. If you love your job, get on with your colleagues and know that similar work is scarce, then it's going to be hard to feel hopeful. Looking on the bright side has to be a good idea, but sometimes there doesn't seem to be a bright side.

By examining the ways in which hopeful people differ from the less hopeful, Rick Snyder uses counselling to help people to increase their levels of hope. He encourages the setting of multiple, specific goals. For example, instead of setting the goal of becoming fitter, he changes it to a goal to exercise twice a week, a goal which is both achievable and tangible. As soon as people begin reaching goals they start to feel more hopeful that they can achieve other goals. Through the setting of appropriate tasks, Chris Peterson believes you can even influence a child's level of optimism. The tasks need to be achievable, but not so easy that the child feels patronised. This approach has been tested systematically by Martin Seligman. He set up a project called the Penn Optimism Program with the specific aim of training children to be more optimistic. Small groups of middle-school children thought to be at risk of depression were invited to attend a twelve-week course during which they were given examples that demonstrated the way that feelings are influenced by thoughts. They were taught that they do have a choice in the way they think about a particular situation. In one scenario they had asked someone to dance, but had been turned down. Instead of jumping to the conclusion that this was because they were unattractive, unlovable and unpopular, they were encouraged to think of other possible reasons why the person might have said no. They were taught not to draw broad conclusions from one incident, nor to blame themselves for everything. However, Seligman was keen that the children should still accept that an event could be bad, although perhaps not as disastrous as they feared. Then he waited to see what happened. When the children were followed up two years later, 22% showed moderate to severe symptoms of depression compared with 44% of a group of children who started out with the same risk of depression but didn't attend the programme.

It's notable that Seligman took care to teach *realistic* thinking rather than simply promoting positive thinking, which if *Pollyanna* is anything to go by can be extremely irritating. In Eleanor H. Porter's novel for children Pollyanna perfected the art of optimism with her 'glad game' with which she always saw the best in any situation. No one she met was allowed to feel unhappy about their situation or receive sympathy. Instead if someone was bedridden she demanded they feel glad that everyone else wasn't bedridden. Maybe I'm particularly heartless, but I can't have been alone in feeling a little glad myself when she finally got run over by a car. I hoped that then she would realise just what she'd put everyone else through. Instead, of course, despite her paralysis, she soon finds things to be glad about and even learns to walk again.

Pollyanna was at the extremes of hopefulness. Most of us would like to think we are a little more realistic in our outlook on life, but it is possible that most of us are overoptimistic and that this helps us to function. If you ask people whether they are more or less likely than average to suffer from a serious illness or accident, *most* people think something bad is unlikely to befall them and of course most people can't be luckier than average. The cliché 'I never thought it would happen to me' is revealing. The future contains uncertainty for everyone, but we choose to hope that for us, everything will be all right.

There is an extraordinary phenomenon, however, which suggests that it might be the depressed people who are realistic, while everyone else is overoptimistic. In these experiments known as contingency studies a person sits in front of a computer and is told to press the space bar whenever they choose. Intermittently a light flashes up on the screen, sometimes in response to the space bar and sometimes at random. At the end of the exercise they have to estimate the percentage of times the light came on as a result of them pressing the space bar. Most people overestimate the control they have over the computer, but depressed people are often spot-on. This is known as depressive realism. This finding is so robust that some hospitals use contingency tasks to assess whether patients have recovered from depression. As they become less accurate at the task, their state of mind is considered to be improving.

This phenomenon has become known as 'sadder but wiser' after the name of the article in which the findings were first published. Are we perhaps programmed to be over-hopeful? Is over-optimism part of a contented life? We know that in all likelihood not everything will go to plan and that as we age many people we know will suffer illness and that in the end everyone we know will die, including our parents, our friends and ourselves. The future looks bleak, but somehow most people are not paralysed with worry and manage to carry on as though everything will be all right, when it won't. Those who cannot are diagnosed with anxiety or depression.

high hopes

If hope can affect our ability to assess risk, then despite the benefits it brings in terms of success and health, perhaps it's not the case that the more hope we feel, the better. Optimistic people can be worse at calculating risk, because they over-estimate the chances of things working in their favour. When he flies, Chris Peterson hopes his pilot is not an optimist. If there's an ice storm and the pilot has to decide whether to press on or turn back, he'd rather the pilot were cautious. There's good evidence that he's right. A study conducted at the University of Otago in New Zealand investigated why the figures for small private plane crashes are always the same – almost every year the number of small planes which come down in New Zealand is four. Through an examination of the accident reports two common causes were found. One is that pilots fly too low over a building either to wave at someone they know on the ground or because they want to show something to their friend in the passenger seat. This is something that pilots know they shouldn't do, but sometimes can't resist. The other reason, however, is more pertinent. Bad weather closes in and, despite their experience, pilots ignore the danger signs and continue their journey, hoping that they will be all right.

In certain situations, however, optimism can lead to better decisions. Dylan Evans set up a computer game at Bath University in which the competitors were all virtual, but some were programmed to make rational judgements and others to behave like optimistic humans. Under certain conditions the optimistic human-like competitors actually did better. This suggests that when failure won't have negative consequences, but might bring a high reward, it is worth being overoptimistic. For example, a little optimism might encourage you to enter a competition. The chances are that you won't win, but all it will cost you is a few minutes of wasted time, and you might win a fantastic holiday. On the other hand if you're standing between two cliffs at 'Soldier's Leap' above a forty-foot drop to the sea below, trying to decide whether to follow in the footsteps of the fleeing soldier by attempting the gap yourself, too much optimism would be fatal.

Sometimes the deliberate employment of pessimism can be useful. Defensive pessimism would be an example. This is where you convince yourself that you are probably going to fail and if you subsequently do, you're cushioned from the disappointment. If you succeed on the other hand, it comes as a nice surprise. Some people find this is an effective method of coping with a situation because in anticipating failure their anxiety is reduced. The psychologist Julie Norem found defensive pessimists actually performed worse if they were prevented from reflecting on their low expectations before a task or if their mood was enhanced beforehand. For them, pessimism seems to work. When people reach old age, Seligman found that people who were realistically pessimistic coped better than those who were overly optimistic. This is not the case with younger people, where it's the optimistic who deal with negative events better, as we saw with the students after September 11th.

Perhaps this is why hope plays an ambiguous role in the myth of Pandora's box. Depending on whose version of the story you read, hope was either the saviour of the human race or yet another negative emotion, waiting to make our lives a misery. Pandora was sent down to earth by the gods and presented to Prometheus' brother, Epimetheus. Unfortunately they sent with her a jar (not a box) full of human ills. She was told on no account to ever open the jar and so naturally she did. Out poured plague, gout, rheumatism, colic, spite, envy and revenge. As she tried to slam the jar shut, hope became stuck under the lip. In some versions of the story hope then slipped out at the last moment, ready to make life difficult for humans by encouraging them to waste their time on untenable enterprises. In other versions hope remained in the jar, left behind to help humankind deal with all the other ills which had been released on the world. Hope remains both positive and negative. Nietzsche called it 'the worst of all evils, for it protracts the torment of man'.

With the help of a large group of students the emotions researcher James Averill came up with an extraordinary 300 metaphors for hope. These included plenty of phrases suggesting that hope can be negative, many of which compared hope to food such as 'hope is a good breakfast but an ill supper', 'he who lives on hope will die fasting' and 'hope is the poor man's bread'. Other phrases suggested that hope deceives us, leading us astray: 'hope is a charlatan who always defrauds us', 'hope is as cheap as despair', 'blinded by hope' and 'the houses that hope builds are castles in the air'.

Hope certainly can present specific problems when people are ill. Despite the benefits of fighting spirit, hope can be a negative force, depending on the particular illness involved. Researchers at Utrecht University found that optimism protects against depression in people with conditions such as multiple sclerosis, which involve huge amounts of uncertainty, but few medical guidelines to follow. However, if you have a disease which requires you to follow certain regimes, such as diabetes, then an optimistic outlook might prevent you from adhering to a doctor's advice.

false hope

At the University of Toronto Janet Polivy is studying false hope syndrome. This occurs when a person sets themselves an unrealistic target, inevitably fails and then feels bad about themselves as a result. She has studied people who make the same new year's resolutions every year, sometimes for ten years in a row. Every year they fail, but it never occurs to them that it's not their willpower that's the problem, but the resolution. Polivy is astonished by the optimism with which people approach yet another diet, always convinced that this time it will work. This shows just how powerful hope can be. Some students started fifteen new diets in a year. The problem was not that they have too much hope *per se*, but that those hopes lack any basis in reality.

Women embarking on a diet were asked to name different amounts of weight loss which would make them feel each of the following: happy, satisfied, disappointed or a failure. By the end of the diet programme women had lost on average thirty-seven pounds (seventeen kilos) which was not bad at all, but they had been so overoptimistic regarding the amount of weight they thought they could lose that most hadn't even reached their 'disappointed' weight. Part of the reason concerns the nature of dieting; after the first few pounds you have to eat even less and exercise even more in order to continue losing weight, so perhaps it's not surprising that it's harder than people expect. However Polivy found the same pattern also occurred when people were trying to change other aspects of themselves. When she studied students whose aim was to take more exercise or to be nicer to people, false hopes were once again present. Their expectations were unrealistic in four ways: they set their targets too high, thought the changes could be made fast, that it would be easy, and that other changes in their lives would follow on from this one change. For example, she found that students thought that if they lost weight they would not only find a new boyfriend but obtain a better job and achieve better grades. Even if they had succeeded in losing a lot of weight they were then hugely disappointed that their whole lives weren't improved. Their high levels of hope were making them unhappy. Even those who had begun visiting the gym twice a week thought they had failed in their bid to get fit because their aim was to go every single day. They did not consider exercising twice a week to be an improvement, but evidence of failure.

Feelings of hope are reinforcing, however. At the start of a new plan, the false hopes make people feel good because they feel in control and this of course encourages them to try again. Polivy even found that when people had just joined an exercise programme they felt taller. Despite people's evident determination to succeed, a quarter of new year's resolutions are abandoned within a week and most people have given up by the end of January; 60% will make the same resolution again next year. Polivy questions why we wait for a new year before making a resolution, but I think it makes sense. Many resolutions concern healthy living and if you celebrate Christmas then any attempt at eating or drinking less or exercising more is destined to fail if you begin on 15th December. Moreover, it may be just a trick of the calendar, but a new year provides the psychological boost of a brand new start.

As with anything, Polivy's data can be looked at in two ways – glass half full or glass half empty. It does show how often people fail, but it also demonstrates the extraordinary determination they possess to make another attempt, even when the odds are that they will fail. The problem is that failure has its costs and each failed attempt reduces self-esteem. However, Polivy's solution is not for people to abandon hope. Instead she says we need to take things one step at a time, carefully analysing what we want to change about ourselves, why that matters and what we can realistically achieve. Once we reach one small goal, we can go on to the next. At her university she set up a realistic weight-loss programme to test her theories. The result was that students set realistic goals, lost weight and felt satisfied with the results.

Sometimes a dose of realism is needed to counteract false hopes. Rachel, whose husband died in the bicycle accident, would find it frustrating when he poured cold water on one of her latest ideas. Now that he's gone she has realised that by dampening her enthusiasm he was getting her to slow down and think things through. She misses that now and says the reason their relationship worked so well was that while she was so optimistic, he was the realist.

Occasionally false hope is exactly what people need. Primo Levi, describing his time as a prisoner in Auschwitz, wrote that rumours would sometimes circulate the camp that the war was to finish in two weeks or that no more people were to be sent to the gas chambers. His friend Alberto had always criticised the people who consoled themselves with these false hopes until the day that his own father was selected for the gas chamber. 'In the space of a few hours Alberto changed.' He started talking of stories he'd heard that the Russians were nearby and that today's selection had been different; instead of choosing people for the gas chamber, they were taking the weakest people to a special convalescent camp. In 1945 the survivors were marched out of the camp and somehow Alberto disappeared. When Primo Levi visited Alberto's mother to tell her what he knew of Alberto's disappearance, she too was comforting herself with false hopes. She believed that he must have been the one person to have escaped from the lines without being shot. A year later she clung to the hope that Alberto might be in a Russian hospital, unable to make contact because he had lost his memory.

The difficulty with hope is to find a way of harnessing its useful aspects, without letting it lead you astray. What we need are dreams which aren't pure fantasy, but that's not easy to gauge. Suppose your thirteen-year-old who possesses a good singing voice decides she wants to be a pop singer. Should she be discouraged because she is likely to end up disappointed or encouraged because a few people are going to be successful singers at the age of twenty, so why not her?

If hope really can have such an effect perhaps we should focus more attention on it, learning strategies for preserving hope when it seems far away. If hope is constant from the age of eight, then early childhood is crucial and that would be the time to learn. It can bring success in so many areas – health, work, exam grades, sport and even longevity. If it can make a bigger difference to how long you live than smoking does, perhaps it is warnings against the dangers of hopelessness that we need.

As with joy there's a somewhat dispiriting unfairness here. If you are already hopeful good things are more likely to happen to reinforce that hope, which leaves little to look forward to for people who already feel hopeless about the future. But perhaps that's the pessimist's way of approaching the subject. Considered the other way around, it means that if hope can somehow be harnessed, a change in luck might be possible.

afterword

At the beginning of the book we left Reg in a German POW camp in 1944, feeling sad and angry, but still retaining hope that the war would soon end so that he could finally meet up with his sweetheart Marjory after five long years of captivity. Meanwhile Marjory had shown great determination to overcome the back and pelvic injuries which had threatened to leave her paralysed. With the help of crutches she was even able to walk to the hospital to have a series of x-rays taken as part of an assessment for the army medical board.

While Marjory lay in the x-ray room that day a nurse rushed in. Everyone must come and listen to the wireless at once. Marjory was furious that she was abandoned, undressed and unable to move, but through the open door she could hear Churchill's voice, and her anger turned to elation as he announced that the war in Europe was over. Soon the POW camp would be liberated and Reg would be on his way home.

For Reg, however, there were still some unpleasant emotions to experience. Before his return to England he was to see at first hand one of the greatest evils of the war. The American forces needed three British soldiers to accompany them as witnesses to Auschwitz and Reg was one of the officers who went. The bodies of those killed by the Nazis were still being burnt and to this day he recalls the smell. 'At the very end of the war one should be happy, but then we saw that.'

With his mind still full of the horrors of Auschwitz, Reg flew back to England only to discover on landing that the other plane carrying POWs from his camp had crashed, killing all on board. They had survived five years of captivity only to die on the journey home. It was the cruellest of twists. Why should he have survived when they didn't? It made it even harder to put his wartime experiences behind him. For decades his nightmares about collapsing tunnels were to persist and his anger took years to diminish; he even found himself shouting at Germans in the street.

On holiday in Malta many years after the war he did meet one German for whom he had nothing but admiration, a German Jew who had escaped to Britain before the war. While he was interned he spent his time creating inventions for British soldiers in POW camps. He had no idea whether they were ever used, but the invention of which he was most proud was the fountain pen which contained a foul smell. All those years after the war he finally met a man who had used it, who had experienced the joy of getting one over on the opposition.

The day after Reg returned to England, he and Marjory went on their long-postponed first date. When they met under the clock at Waterloo station, they both showed the physical effects of the last five years. The young officer was still good-looking with smooth brown hair and a neat moustache, but after five years of meagre meals he was thin and undernourished. The lovely dancer still had fine cheekbones and a mass of blonde hair, but there were no longer high heels at the end of her long legs. Now she wore a leg iron and walked with sticks. Their date did not lack style, however – a taxi to the Savoy Hotel on the Thames, for lunch accompanied by a magnum of champagne. Just a few weeks later they got engaged and they married on VJ day. They have been married now for almost sixty years. My great aunt and uncle's story is one of anger, disgust, fear, guilt, joy and hope, but above all, love.

Each of these emotions had its place in their story, but, as we've seen, emotions cannot simply be divided into positive and negative. When the circumstances are right they all have their place and are wiser than we might think. Firstly they provide us with information. If an activity makes us feel joyful we can seek to do more of it. If we feel angry about something somebody has said to us, we can look carefully at why we feel so upset. Identifying the emotion gives us information. We might think we feel angry, but are we in fact sad because we've lost the relationship we had with that person? Or are we jealous that they're focusing their attention elsewhere, or guilty because we know we're to blame or even afraid of what we might lose? Even our bodies can provide clues as to our emotional state and it's extraordinary that our abilities at detecting those clues might affect the way we feel.

Emotions are often regarded as a problem, but the difficulty is not that they possess us, forcing us to behave irrationally.

The problem is that the right emotion does not always appear at the right moment. Emotions play a crucial part in communication. There is a time for anger and a time for fear. We need those emotions, but it is the timing that is so hard.

In this book I've drawn together the work of neuroscientists, psychologists and biologists. For decades researchers have been divided over which emotions are evolutionary adaptations or social constructions. In order to allow both theories to coexist some have suggested that we evolved to feel the basic emotions like fear and anger, while societies developed the complex emotions like love and guilt. But even love, guilt and jealousy can be useful for survival, as I've illustrated. It seems more sensible to follow the theory developed by the philosopher Jesse Prinz from the University of North Carolina, which proposes that every emotion can be useful to survival, as well as being influenced by society. We may have evolved to feel fear, but as we grow up we learn to take context into account; if you live in London and see a lion two feet away from you, then the chances are that it's in a cage at the zoo and you needn't be afraid. If you work in a Kenyan game reserve and see a lion this close you would respond rather differently. Likewise, each emotion can be instant and primitive at some moments yet involve sophisticated reasoning at others. You might suddenly feel afraid when, alone in a house at night, you hear a noise downstairs, or you might have a slow build-up of fear as you watch your partner's behaviour over many months and gradually realise that they are about to leave you.

Our culture shapes each of our emotions as we grow up, with the result that societies have different objects of fear and disgust, different situations where the expression of emotion such as anger might be appropriate, as well as different attitudes towards acting on your emotions, such as following your heart if you are in love.

Whichever way we approach the emotions, understanding them can reduce their mystery. Emotions exist to help us, providing us with a quick way of assessing and responding to a situation whilst keeping us away from danger, whether it be a potential attacker or some poisonous food. Through feelings like joy and hope we are constantly encouraged to seek out new possibilities and to find moments of happiness, even in the grimmest situations. Mark Henderson, who spent three months as a hostage in the Colombian jungle, wrote in his diary that one night after three days of a diet of lentils and beans, he and his fellow hostages laughed themselves to sleep at their own flatulence. Our capacity to search for moments of joy during the most horrendous situations is extraordinary.

If we consider our emotions at the level of brain chemistry, it is clear that outside events can affect the levels of different neurotransmitters in the brain. We can even deliberately alter those chemicals ourselves by something as simple as a brisk walk or listening to one happy track from a CD. Viewed at a cognitive level we can influence our emotions by changing the way we look at things. Can we really infer from one mistake that we are hopeless? Although none of this sounds very mysterious, emotions have somehow come to occupy a position of mystery. Rather than tools, they tend to be seen as feelings which overtake us. In fact they are ours to exploit.

2550

bibliography

Due to the quantity of research on the science of emotions this list is not exhaustive, but I have included the references I feel are the most useful. With apologies to second and third authors, where papers have multiple authors I have saved space and trees by only including the first.

the science of emotions: general references covering several emotions

Cornelius, R. R. (1996) *The Science of Emotion*. New Jersey: Prentice-Hall.

Damasio, A. (2000) *The Feeling Of What Happens – Body, Emotion And The Making Of Consciousness*. London: Vintage.

Darwin, C. (1872/1999) *The Expression of the Emotions in Man and Animals*. London: HarperCollins.

Davidson, R. (2003) *Handbook of Affective Sciences*. New York: Oxford University Press.

Ekman, P. (2003) *Emotions Revealed – Understanding Faces and Feelings*. London: Weidenfeld & Nicolson.

Evans, D. (2001) *Emotion – The Science of Sentiment.* Oxford: Oxford University Press.

Frijda, N. H. (1986) *The Emotions.* Cambridge: Cambridge University Press.

Goleman, D. (1996) *Emotional Intelligence.* London: Bloomsbury.

Harré, R. (1996) *The Emotions: social, cultural and biological dimensions.* London: Sage.

Izard, C. E. (1991) *The Psychology of Emotions.* New York: Plenum Press.

James, W. (1890) 'The Principles of Psychology'. In Jenkins, J. M. et al. (eds) (1998) *Human Emotions: A Reader.* Malden, Massachusetts: Blackwell Publishers.

Lane, R. D. et al. (1997) 'Neuroanatomical Correlates of Happiness, Sadness, and Disgust'. *American Journal of Psychiatry* 154, 926–933.

LeDoux, J. (2002) *Synaptic Self: How our brains become who we are.* London: Macmillan.

Lewis, M. & Haviland-Jones, J. M. (eds) (2000) *Handbook of Emotions.* New York: The Guilford Press.

Masson, J. & McCarthy, S. (2000) *When Elephants Weep.* London: Vintage.

Nussbaum, M. C. (2001) *Upheavals of Thought: the intelligence of emotions.* Cambridge: Cambridge University Press.

Oatley, Keith & Jenkins, J. M. (1996) *Understanding Emotions.* Oxford: Blackwell Publishers.

Parkinson, B. (1995) *Ideas and Realities of Emotion.* London: Routledge.

Ratey, J. (2001) *A User's Guide to the Brain.* London: Little, Brown & Co.

Rolls, E. T. (1999) *The Brain and Emotion.* Oxford: Oxford University Press.

Strongman, K. T. (1996) *The Psychology of Emotion*. Chichester: John Wiley.

development of emotions in children

Berryman, J. C. et al. (1991) *Developmental Psychology and You*. Leicester: British Psychological Society.
Darwin, C. (1877) 'A biographical sketch of an infant'. *Mind Quarterly Review of Psychology and Philosophy* 7 (July), 285–294.
Draghi-Lorenz, R. (2003) *Young Infants are Capable of 'Non-Basic' Emotions*. PhD thesis.
Harris, P. (1989) *Children & Emotion: The Development of Psychological Understanding*. Oxford: Blackwell.
Hobson, P. (2002) *The Cradle of Thought*. London: Macmillan.
Izard, C. E. & Malatesta, C. Z. (1987) 'Perspectives on Emotional Development'. In Osofsky, J. D. (ed.) *Handbook of Infant Development*. New York: Wiley.
Lewis, M. (2000) 'The Emergence of Human Emotion'. In Lewis, M. & Haviland-Jones, J. M. (See general list above.)
Malatesta, C. Z. et al. (1989) 'The Development of Emotional Expression During the First Two Years of Life'. *Monographs of the Society for Research in Child Development* 54(1–2).

joy

Alexander, B. et al. (1998) 'Adult, infant and animal addiction'. In Peele, S. (ed.) *The Meaning of Addiction*. San Francisco: Jossey-Bass.
Apter, M. (2001) *Motivational Styles in Everyday Life*. Washington DC: American Psychological Association.
Argyle, M. (1987) *The Psychology of Happiness*. London: Methuen.

Damasio, A. (2003) *Looking For Spinoza*. London: William Heinemann.

Fernandez-Dols, J. & Ruiz-Belda, M. (1995) 'Are Smiles a Sign of Happiness?: Gold Medal Winners at the Olympic Games'. *Journal of Personality & Social Psychology* 69, 1113–1119.

Fredrickson, B. L. (1998) 'What Good are Positive Emotions?' *Review of General Psychology* 2, 300–319.

Fredrickson, B. L. (2003) 'The Value of Positive Emotions'. *American Scientist* 91, 330–335.

Isen, A. M. (2000) 'Positive Affect and Decision Making'. In Lewis, M. & Haviland-Jones, J. M. (See general list above.)

Lewis, M. (2000) 'The Emergence of Human Emotion'. In Lewis, M. & Haviland-Jones, J. M. (See general list above.)

Morgan, W. (ed.) (1997) *Physical Activity and Mental Health*. Philadelphia: Taylor & Francis.

Noble, E. P. (2000) 'Addiction and its reward process through polymorphisms of the D2 dopamine receptor gene: A review'. *European Psychiatry* 15(2), 7–89.

Olds, J. & Milner, P. (1954) 'Positive reinforcement produced by electrical stimulation of sepatal areas and other regions of rat brains'. *Journal of Comparative and Physiological Psychology* 47, 419–427.

Panksepp, J. & Gordon, N. (2003) 'The instinctual basis of human affect: affective imaging of laughter and crying'. *Consciousness & Emotion* 4(2), 197–205.

Phillips, W. T. et al. (2001) 'The Effects of Physical Activity on Physical and Psychological Health'. In Baum, A. et al. (eds) *Handbook of Health Psychology*. New Jersey: Lawrence Erlbaum.

Raleigh, M. J. et al. (1991) 'Serotonergic mechanisms promote dominant acquisition in adult male vervet monkeys'. *Brain Research* 559, 181–190.

Rankin, A. M. & Philip, P. J. (1963) 'An Epidemic of Laughing in the Bukoba District of Tanganyika'. *Central African Journal of Medicine* 12(9), 167–170.

Steinberg, H. & Sykes, E. A. (1985) 'Introduction to symposium on endorphins and behavioural processes: review of literature on endorphins and exercise'. *Pharmacology, Biochemistry & Behavior* 23, 857–862.

Steinberg, H. et al. (1997) 'Exercise enhances creativity independently of mood'. *British Journal of Sports Medicine* 31, 240–245.

Steinberg, H. et al. (1999) 'Weekly Exercise Consistently Reinstates Positive Mood'. *Psychologie in Osterreich* 4–5, 265–274.

sadness

Alloy, L. B. & Abramson, L. Y. (1979) 'Judgement of contingency in depressed and non-depressed subjects: Sadder but wiser?' *Journal of Experimental Psychology: General* 108, 443–479.

Arborelius, L. et al. (1999) 'The role of corticotrophin-releasing factor in depression and anxiety disorders'. *Journal of Epidemiology* 160, 1–12.

Barr-Zisowitz, C. (2000) '"Sadness" – Is There Such a Thing?' In Lewis, M. & Haviland-Jones, J. M. (See general list above.)

Becht, M. C. et al. (2001) 'Crying Across Countries'. In Vingerhoets, A. J. J. M. & Cornelius, R. R. (eds), *Adult Crying: A Biopsychosocial Approach*. Hove: Brunner-Routledge.

Brammer, G. L. et al. (1994) 'Neurotransmitters and social status'. In Ellis, L. (ed.) *Social Stratification and Socioeconomic Quality Vol.2: Reproductive and Interpersonal Aspects of Dominance and Status*. Westport, CT: Praeger.

Brown, G. W. & Harris, T. (1978) *Social Origins of Depression.* London: Tavistock.

Frey, W. H. (1986) *Crying: The Mystery of Tears.* USA: Winston Press.

Izard, C. E. (1991) *The Psychology of Emotions.* New York: Plenum Press.

Matthews, K. A. et al. (2000) 'Does Socioeconomic status relate to central serotonergic responsivity in healthy adults?' *Psychosomatic Medicine* 62, 231–237.

Nemeroff, C. B. (1998) 'The neurobiology of depression'. *Scientific American* 278, 42–49.

Nesse, R. M. (2001) 'Motivation and Melancholy: A Darwinian Perspective'. Nebraska Symposium on Motivation.

Stevens, A. & Price, J. (1996) *Evolutionary Psychiatry.* London: Routledge.

Tse, W. A. & Bond, A. J. (2002) 'Serotonergic intervention affects both social dominance and affiliative behaviour'. *Psychopharmacology* 161(3), 324–330.

Vingerhoets, A. J. J. M. & Cornelius, R. R. (eds) (2001) *Adult Crying: A Biopsychosocial Approach.* Hove: Brunner-Routledge.

disgust

Arens, W. (1979) *The Man-Eating Myth – Anthropology and Anthropophagy.* Oxford: Oxford University Press.

Brown, P. & Tuzin, D. (eds) (1983) *The Ethnography of Cannibalism.* Washington DC: Society for Psychological Anthropology.

Conklin, B. A. (2001) *Consuming Grief: Compassionate Cannibalism in an Amazonian Society.* Austin: University of Texas Press.

De Jong, P. J. et al. (2002) 'Disgust and disgust sensitivity in spider phobia: Facial EMG in response to spider and oral disgust imagery'. *Journal of Anxiety Disorders* 16, 477–493.

Douglas, M. (1966) *Purity and Danger*. London: Ark Paperbacks.

Graham, H. (1899/2000) *Ruthless Rhymes for Heartless Homes & More Ruthless Rhymes*. New York: Dover Publications.

McNally, R. J. (2002) 'Disgust has arrived'. *Journal of Anxiety Disorders* 16, 561–566.

Miller, W. I. (1997) *The Anatomy of Disgust*. Cambridge, Massachusetts: Harvard University Press.

Phillips, M. L. et al. (1997) 'A specific neural substrate for perception of facial expressions of disgust'. *Nature* 389, 495–498.

Phillips, M. L. et al. (1998) 'Disgust – the forgotten emotion of psychiatry'. *British Journal of Psychiatry* 172, 373–375.

Phillips, M. L. et al. (2000) 'A differential neural response in obsessive-compulsive disorder patients with washing compared with checking symptoms to disgust'. *Psychological Medicine* 30, 1037–1050.

Rozin, P., Haidt, J. & McCauley, C. R. (2000) 'Disgust'. In Lewis, M. & Haviland-Jones, J. M. (See general list above.)

anger

Aldridge, S. (2001) *Seeing Red and Feeling Blue: The New Understanding of Mood and Emotion*. London: Arrow Books.

Averill, J. R. (1982) *Anger and Aggression: An Essay on Emotion*. New York: Springer-Verlag.

Basore, J. W. (trans.) (1935) *Seneca: Moral Essays*. London: Heinemann.

Briner, R. B. (1999) 'The Neglect and Importance of Emotion at Work'. *European Journal of Work and Organizational Psychology* 8, 323–346.

Briner, R. B. & Totterdell, P. (2002) 'The experience, expression and management of emotion at work'. In Warr, P. (ed.) *Psychology at Work.* Fifth Edition. London: Penguin.

Dryden, W. (1996) *Overcoming Anger.* London: Sheldon Press.

Gammie, S. C. & Nelson, R. J. (1999) 'Maternal aggression is reduced in neuronal nitric oxide synthase-deficient mice'. *Journal of Neuroscience* 19, 8027–8035.

Gold, A. & Johnston, D. W. (1991) 'Anger, hypertension and heart disease'. *Current Developments in Health Psychology.* London: Harwood Academic Press.

Hansen, C. H. & Hansen, R. S. (1994) 'Emotion: Attention and Facial Efference'. In Niedenthal, Paula M. & Kitayama, S. (eds) *The Heart's Eye: Emotional Influences in Perception and Attention.* San Diego: Academic Press.

Keinan, G. et al. (1992) 'Anger in or out, which is healthier? An attempt to reconcile inconsistent findings'. *Psychology and Health* 7, 83–98.

McDermott, M. R. (1997) 'Looking for "Fido": evaluating research on expressed anger's role in the pathogenesis of coronary artery disease'. *Health Psychology Update* 27, 34–36.

McDermott, M. R. et al. (2001) 'Components of the anger-hostility complex as risk factors for coronary artery disease severity: a multi-measure study'. *Journal of Health Psychology* 6, 309–319.

Martin, P. (1998) *The Sickening Mind: Brain, Behaviour, Immunity & Disease.* London: Flamingo.

Parkinson, B. (2001) 'Anger on and off the road'. *British Journal of Psychology* 92, 507–526.

Ramsey, J. M. C. et al. (2001) 'Components of anger-hostility complex and symptom reporting in patients with coronary artery disease: a multi-measure study'. *Journal of Health Psychology* 6, 713–729.

Tavris, C. (1989) *Anger – The Misunderstood Emotion.* New York: Simon and Schuster.

fear

Ainsworth, M. et al. (1978) *Patterns of Attachment: A Psychological Study of the Strange Situation.* New Jersey: Lawrence Erlbaum.

Ax, A. F. (1953) 'The physiological differentiation between fear and anger in humans'. *Psychosomatic Medicine* XV(5), 433–442.

Critchley, H. D. et al. (2004) 'Neural systems supporting interoceptive awareness'. *Nature Neuroscience* 7, 189–195.

Ekman, P. et al. (1985) 'Is the startle reaction an emotion?' *Journal of Personality and Social Psychology* 49, 1416–1426.

Field, A. P. et al. (2001) 'Who's afraid of the big bad wolf?: a prospective paradigm to test Rachman's indirect pathways in children'. *Behaviour Research and Therapy* 39, 1259–1276.

Field, A. P. et al. (2002) 'Fear information and social phobic beliefs in children: a prospective paradigm and preliminary results'. *Behaviour Research and Therapy* 41, 113–123.

Field, A. P. & Lawson, J. (2003) 'Fear information and the development of fears during childhood: effects on implicit fear responses and behavioural avoidance'. *Behaviour Research and Therapy* 41, 1277–1293.

Flegr, J. et al. (2002) 'Increased risk of traffic accidents in subjects with latent toxoplasmosis: a retrospective case-control study'. *BMC Infectious Diseases* 2(11), 1–13.

Jones, G. E. (1994) 'Perceptions of visceral sensations: a review of recent findings, methodologies and future directions'. In Jennings, J. R. et al. (eds) *Advances in Psychophysiology* 5. London: Jessica Kingsley Publishers.

LeDoux, J. (1998) *The Emotional Brain.* London: Simon & Schuster.

Manyande, A. & Salmon, P. (1998) 'Effects of pre-operative relaxation on post-operative analgesia; immediate increase and delayed reduction'. *British Journal of Health Psychology* 3, 215–224.

Screech, M. A. (trans.) (1993) Michel de Montaigne: *The Essays: A Selection.* London: Penguin.

Shephard, B. (2002) *A War of Nerves: Soldiers and Psychiatrists 1914–1994.* London: Pimlico.

Simons, R. C. (1996) *Boo! Culture, Experience and the Startle Reflex.* Oxford: Oxford University Press.

Skirrow, P. et al. (2001) 'Intensive care – easing the trauma'. *The Psychologist* 14(12), 640–642.

Tucker, N. (1994) 'Suffer Little Children'. *The Times Higher,* 18th November, 18–20.

Waddell, J., Heldt, S. & Falls, W. A. (2003) 'Posttraining lesion of the superior colliculus interferes with feature-negative discrimination of fear-potentiated startle'. *Behavioural Brain Research* 142, 115–124.

jealousy

Brown, W. M. & Moore, G. (2004) 'Fluctuating asymmetry and romantic jealousy'. *Evolution and Human Behavior* 24, 113–117.

Buss, D. (2000) *The Dangerous Passion – Why Jealousy is as Necessary as Love and Sex.* New York: Free Press.

Delgado, A. R. & Bond, R. A. (1993) 'Attenuating the attribution of responsibility: the lay perception of jealousy as a motive for wife battery'. *Journal of Applied Social Psychology* 23, 1337–1356.

DeSteno, D. & Salovey, P. (1996) 'Jealousy and the characteristics of one's rival: a self-evaluation maintenance perspective'. *Personality and Social Psychology Bulletin* 22(9), 920–932.

DeSteno, D. et al. (2002) 'Sex differences in jealousy: evolutionary mechanism or artefact of measurement?' *Journal of Personality and Social Psychology* 83(5), 1103–1116.

Dijkstra, P. & Buunk, B. P. (1998) 'Jealousy as a function of rival characteristics: an evolutionary perspective'. *Personality and Social Psychology Bulletin* 24(11), 1158–1166.

Dolan, M. & Bishay, N. R. (1996) 'The role of the sexual behavior/attractiveness schema in morbid jealousy'. *Journal of Cognitive Psychotherapy* 10(1), 41–61.

Draghi-Lorenz, R. (2003) *Young Infants are Capable of 'Non-Basic' Emotions.* PhD thesis.

Dunn, J. (1983) 'Sibling Relationships in Early Childhood'. *Child Development* 54, 787–811.

Dunn, J. & Kendrick, C. (1982) *Siblings.* London: Grant McIntyre.

Ellis, A. (1977) 'Rational and Irrational Jealousy'. In Clanton, G. & Smith, L. G. (eds) *Jealousy.* New Jersey: Prentice-Hall.

Gilmartin, B. G. (1977) 'Jealousy Amongst the Swingers'. In Clanton, G. & Smith, L. G. (eds) *Jealousy.* New Jersey: Prentice-Hall.

Harris, C. R. (2000) 'Psychophysiological responses to imagined infidelity: the specific innate modular view of jealousy reconsidered'. *Journal of Personality and Social Psychology* 78(6), 1082–1091.

Harris, C. R. (2003) 'A review of sex differences in sexual jealousy, including self-report data, psychophysiological responses, interpersonal violence, and morbid jealousy'. *Personality and Social Psychology Review* 7(2), 102–128.

Mayr, E. (2002) *What Evolution Is.* London: Phoenix.
Mead, M. (1931/1977) 'Jealousy: Primitive and Civilised'. In Clanton, G. & Smith, L. G. (eds) *Jealousy.* New Jersey: Prentice-Hall.
Parks, C. D. et al. (2002) 'The effects of envy on reciprocation in a social dilemma'. *Personality and Social Psychology Bulletin* 28(4), 509–520.
Pines, A. & Aronson, E. (1983) 'Antecedents, correlates, and consequences of sexual jealousy'. *Journal of Personality* 51(1), 108–136.
Rose, H. & Rose, S. (2001) *Alas Poor Darwin – Arguments Against Evolutionary Psychology.* London: Vintage.
Salovey, P. & Rodin, J. (1986) 'The differentiation of social-comparison jealousy and romantic jealousy'. *Journal of Personality and Social Psychology* 50(6), 1100–1112.
Smith, R. H. et al. (1988) 'Envy and jealousy: semantic problems and experiential distinctions'. *Personality and Social Psychology Bulletin* 14(2), 401–409.
Sokoloff, B. (1948) *Jealousy – a psychological study.* London: Caroll & Nicholson.
Stenner, P. & Rogers, R. S. (1998) 'Jealousy as a manifold of divergent understandings: a Q methodological investigation'. *European Journal of Social Psychology* 28, 71–94.
Tarrier, N. et al. (1990) 'Morbid jealousy: a review and cognitive-behavioural formulation'. *British Journal of Psychiatry* 157, 319–326.
Zizzo, D. J. & Oswald, A. (2001) 'Are people willing to pay to reduce others' incomes?' *Annales d'Economie et de Statistique* 63–64, 39–62.

love

Ainsworth, M. et al. (1978) *Patterns of Attachment: A Psychological Study of the Strange Situation*. New Jersey: Lawrence Erlbaum.

Andreae, S. (1998) *Anatomy of Desire: The Science and Psychology of Sex, Love and Marriage*. London: Little, Brown & Co.

Aron, A. (1988) 'The matching hypothesis reconsidered again: comments on Kalick and Hamilton'. *Journal of Personality and Social Psychology* 54(3), 441–446.

Aron, A. et al. (1995) 'Falling in love: prospective studies of self-concept change'. *Journal of Personality and Social Psychology* 69(6), 1102–1112.

Aron, A. et al. (1998) 'Motivations for unreciprocated love'. *Personality and Social Psychology Bulletin* 24(8), 787–796.

Bartels, A. & Zeki, S. (2000) 'The neural basis of romantic love'. *NeuroReport* 11(17), 3829–3834.

Ben-Ari, E. T. (1998) 'Pheromones: what's in a name?' *Bioscience* 48, 505–511.

Dutton, D. & Aron, A. (1974) 'Some evidence for heightened sexual attraction under conditions of high anxiety'. *Journal of Personality and Social Psychology* 30, 510–517.

Fisher, H. E. (1992) *Anatomy of Love*. New York: W. W. Norton & Co.

Gangestad, S. W. et al. (1994) 'Facial attractiveness, developmental stability and fluctuating asymmetry'. *Ethology and Sociobiology* 15, 73–85.

Hatfield, E. & Rapson, R. L. (1993) *Love, Sex and Intimacy: Their Psychology, Biology, and History*. New York: HarperCollins.

Hatfield, E. & Rapson, R. L. (1996) *Love and Sex: Cross-Cultural Perspectives*. Needham Heights, Massachusetts: Allyn & Bacon.

Jankowiak, W. R. & Fischer, E. F. (1992) 'A cross-cultural perspective on romantic love'. In Jenkins, J. M. et al. (1998) *Human Emotions: A Reader.* Malden, Massachusetts: Blackwell Publishers.

Kohl, J. V. et al. (2001) 'Human pheromones; integrating neuroendocrinology and ethology'. *Neuroendocrinology Letters* 22, 309–321.

Liebowitz, M. R. (1983) *The Chemistry of Love.* Boston: Little, Brown.

Miller, G. (2001) *The Mating Mind.* London: Vintage.

Perrett, D. I. et al. (1998) 'Effects of sexual dimorphism on facial attractiveness'. *Nature* 394, 884–887.

Schachter, S. & Singer, J. (1962) 'Cognitive, social and physiological determinants of emotional state'. *Psychological Review* 69, 379–399.

Scholey, A. et al. (1999) 'The effects of exposure to human pheromones on mood and sexual attraction'. *Proceedings of the British Psychological Society* 7(1), 77.

Sergeant, M. J. T. (2003) 'Review of the scent of Eros: mysteries of odour'. In *Human Sexuality* by Kohl, J. V. & Francouer, R. T. *Human Nature Review* 3, 284–288.

Shaver, P. et al. (1988) 'Love as attachment: the integration of three behavioural systems'. In Sternberg, R. J. & Barnes, M. L. (eds) New Haven: Yale University Press.

Solomon, R. C. (1988) *About Love.* New York: Simon & Schuster.

Stendhal (1830/1975) *Love.* London: Penguin.

Sternberg, R. & Barnes, M. L. (eds) (1988) *The Psychology of Love.* New Haven: Yale University Press.

Wedekind, C. et al. (2001) 'MHC-dependent mate preferences in humans'. *Nature* 260, 245–249.

White, G. L. et al. (1981) 'Passionate love and the

misattribution of arousal'. *Journal of Personality and Social Psychology* 41, 56–62.

guilt

Anissimov, M. (trans.) (1998) Primo Levi: *Tragedy of an Optimist*. London: Aurum.

Blacker, R. S. (2000) '"It isn't fair": post-operative depression and other manifestations of survivor guilt'. *General Hospital Psychiatry* 22(1), 43–48.

De Jong, P. J. et al. (2002) 'Blushing after a moral transgression in a prisoner's dilemma game: appeasing or revealing?' *European Journal of Social Psychology* 32, 627–644.

Draghi-Lorenz, R. (2003) *Young Infants are Capable of 'Non-Basic' Emotions*. PhD thesis.

Hull, A. M. et al. (2002) 'Survivors of the Piper Alpha oil platform disaster: long-term follow-up study'. *British Journal of Psychiatry* 181, 433–438.

Levi, P. (1988) *The Drowned and the Saved*. London: Abacus.

Lykken, D. T. (1998) *A Tremor in the Blood: Uses and Abuses of the Lie Detector*. New York: Plenum Press.

Morris, H. (1987) 'Nonmoral guilt'. In Schoeman, F. (ed.) *Responsibility, Character and the Emotions*. Cambridge: Cambridge University Press.

O'Connor, L. E. et al. (2000) 'Survivor guilt, submissive behaviour and evolutionary theory: the down-side of winning in social comparison'. *British Journal of Medical Psychology* 73, 519–530.

Pavlidis, J. et al. (2002) 'Seeing through the face of deception'. *Nature* 415, 35.

Rilling, J. K. et al. (2002) 'A neural basis for social cooperation'. *Neuron* 35, 395–405.

Williams, T. (1988) 'Diagnosis and treatment of survivor guilt'. In Wilson, J. P. et al. (eds) *Human Adaptation to Extreme Stress*. New York: Plenum Press.

Wollheim, Richard (1999) *On the Emotions*. New Haven: Yale University Press.

hope

Alloy, L. B. & Abramson, L. Y. (1979) 'Judgement of contingency in depressed and non-depressed subjects: Sadder but wiser?' *Journal of Experimental Psychology: General* 108, 443–479.

Averill, J. R. et al. (1990) *The Rules of Hope*. New York: Springer-Verlag.

Chang, E. C. (ed.) (2001) *Optimism and Pessimism: Implications for Theory, Research and Practice*. Washington DC: American Psychological Association.

Danner, D. D. et al. (2001) 'Positive emotions in early life and longevity: findings from the nun study'. *Journal of Personality and Social Psychology* 80(5), 804–813.

Evans, D. et al. (2003) 'The evolution of optimism: a multi-agent based model of adaptive bias in human judgement'. *Proceedings of the AISB03 Symposium on Scientific Methods for the Analysis of Agent–Environment Interaction*, University of Wales, pp.20–25.

Evans, D. (2003) *Placebo: The Belief Effect*. London: HarperCollins.

Fredrickson, B. L. et al. (2003) 'What good are positive emotions in crises?: A prospective study of resilience and emotions following the terrorist attacks on the United States on September 11th, 2001'. *Journal of Personality and Social Psychology* 84(2), 365–376.

Greer, S. et al. (1991) 'Psychological response to cancer and survival'. *Psychological Medicine* 21, 43–49.

Langer, E. J. & Rodin, J. (1976) 'The effects of choice and enhanced personal responsibility for the aged: A field experiment in an institutional setting'. *Journal of Personality and Social Psychology* 34, 191–198.

Norem, J. K. (1986) 'Defensive pessimism: Harnessing anxiety as motivation'. *Journal of Personality and Social Psychology* 51(6), 1208–1217.

Peterson, C. (2000) 'The future of optimism'. *American Psychologist* 55(1), 44–55.

Polivy, J. & Herman, C. P. (2000). 'The False Hope Syndrome: Unfulfilled expectations of self-change'. *Current Directions in Psychological Science* 9, 128–131.

Richter, C. P. (1957) 'On the phenomenon of unexplained sudden death in animals and man'. *Psychosomatic Medicine* 19, 191–198.

Seligman, M. E. P. (2002) *Authentic Happiness*. London: Nicholas Brealey.

Shephard, B. (2002) *A War of Nerves: Soldiers and Psychiatrists 1914–1994*. London: Pimlico.

Snowdon, D. (2001) *Aging with Grace: What the Nun Study Teaches Us about Leading Longer, Healthier and More Meaningful Lives*. London: Fourth Estate.

Snyder, C. R. (ed.) (2000) *Handbook of Hope: Theory, Measures and Applications*. San Diego, CA: Academic Press.

Tiger, L. (1979) *Optimism: the Biology of Hope*. London: Secker & Warburg.

Wiseman, R. (2003) 'The Luck Factor'. *Skeptical Inquirer* 27(3).

index

a

b

c

d

g

h

i

j

k

l

m

n

o

p

q

r

s

t

v

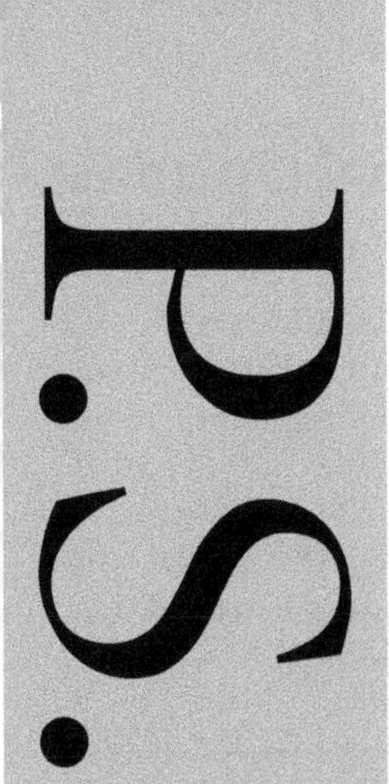

Ideas, interviews & features . . .

About the author

About the book

Read on

An Emotional Optimist

Louise Tucker talks to Claudia Hammond

How did you get into broadcasting?
I was lucky because I always knew from very early on what I wanted to do, which I think is three-quarters of the battle. According to my parents we went to a children's book festival where I met Roald Dahl. He asked, 'Do you know what you want to do?' and I said, 'I want to work in broadcasting.' Obviously thousands of people want to do that now, but then it wasn't seen as a serious job. When I told the careers teacher at school she said, 'Do your parents know about this?' I remember looking up radio and TV in the careers index box and all I could find was TV aerial erector, which I didn't think I'd be very good at.

Why did you then choose to study psychology?
When I was about fourteen I worked in hospital radio in Bedford, where I presented a Sunday night request programme badly. I went round the wards beforehand collecting requests and ended up going earlier and earlier so that I could spend longer and longer talking to patients. Everyone, but particularly the old ladies, would have their requests ready (usually 'The Old Rugged Cross'), not because they wanted to hear them – I'm sure nobody ever listened to the programme at all – but a request was an excuse to chat to me for longer and tell me about their illnesses. I noticed how desperate people were to talk about what was wrong with them; they'd ask me questions that they didn't like to ask the doctor so I became really interested in health psychology. Some people

said, 'Why don't you go and do media studies?' but I thought I'd learn about that working in the industry and wanted to do something different.

How did your family influence your career, if at all?

My mum wanted to be an artist and my dad wanted to work for the RSPB but both their parents persuaded them to go to law school. So they met there, but both hated it and left. Within a few years my dad was doing what he wanted. Then much more recently my mother trained as a botanical artist and now exhibits round the world and wins prizes. As a result of being forced into careers they didn't want, they were determined that I should choose for myself.

There is probably more of an influence book-wise because my dad has written about twenty books about birds, wildlife and wildlife art and I remember being very impressed when his first one arrived when I was about seven. It was a paperback for children called *Looking at Wildlife* and I said to him, 'When are you going to write a proper book?' What I really wanted was a story because that was a *real* book.

What has psychology taught you about yourself?

Loads of people go and study psychology because they think it will teach them about themselves and are usually very disappointed. It's not a self-help subject. There are things you can observe but for ▶

'Loads of people study psychology because they think it will teach them about themselves and are usually very disappointed. It's not a self-help subject.'

Author photo by Niall McDiarmid

LIFE
at a Glance

BORN

Sandy, Bedfordshire, 1971

EDUCATED

Local primary and junior schools, then Dame Alice Harpur School in Bedford; Sussex University; Surrey University

LIVES

London

PAST JOBS

Bakery assistant, factory worker, waitress, local radio newsroom assistant

CURRENT JOBS

Radio presenter, psychology lecturer, journalist and columnist

An Emotional Optimist *(continued)*

◀ the most part psychology today is about the study of large numbers of people; it's not about introspection. But writing the book has made me notice some things about my emotions: for example, when I'm speaking at literary festivals I start to feel sick and my heart rate changes and I think, *why am I putting myself through this, I haven't got to do this, why didn't I just say no?* even though I love it whilst I'm doing it and am happy afterwards. But then I try to put my knowledge into practice, and say to myself, *well, last time I really enjoyed it and it was fine, so what makes you think it's going to go disastrously this time?*

Also if I'm upset about something I ask myself, *what exactly am I upset about, what is it I'm feeling?* because noticing the exact emotions can give you information about what's going on. Finally, in the book I mention that a guy I met said you can top up your happiness levels just by going for a ten-minute walk twice a day and so I do at least that every day, especially when I'm working from home, and I think that does have an effect.

Are you an optimist or a pessimist?

I tend to be quite optimistic about the future. There's a bit in the book about big and little optimism and I'm much better at the big optimism; I think everything will turn out all right in the future. However, if I'm waiting for a bus I'm convinced it won't come, even if it usually does, but that's because I live in London. My husband has very good tube and bus luck: he's never seen a tube information board in double figures so when he turns up the train comes immediately. He's a bit upset if he has to wait two minutes; that's a long time to him. He's a good talisman to have with you.

Why is positive psychology so fashionable now?
Partly because there's this $100,000 prize for the best experiment in the field and partly because it's a rebalancing of the fact that negative emotions predominated and the positive ones were ignored. It is more important to study phobias, anxiety and depression because those are problems, whereas looking for the positive seems like a bonus. It also fits in to the self-help trend. I don't have a lot of time for some of the self-help out there, but Seligman knows what he is talking about. With most of it people don't know what they're getting, whereas with someone like Seligman it's based on research and it's good self-help. Some of the other books promise to change your life quickly and easily but if there really were easy answers we'd have found them.

‘Some books promise to change your life quickly and easily but if there really were easy answers we'd have found them.’

Happiness, according to Seligman and other positive psychologists, is the result of a solid interaction with yourself and the world around you, rather than the immediacy of instant material gratification? Do you agree?
Yes, I think that's definitely true. You can go round getting little pleasures like chocolate and so on but what really sets the happy people apart is that they are happy with ▶

(in no particular order):

Fiction

Anna Karenina
Leo Tolstoy

Unless
Carol Shields

The Comforts of Madness
Paul Sayer

The L-Shaped Room
Lynne Reid Banks

A Suitable Boy
Vikram Seth

Love in the Time of Cholera
Gabriel García Márquez

A Spell of Winter
Helen Dunmore

The Corrections
Jonathan Franzen

The British Museum is Falling Down
David Lodge

The Men Who Loved Evelyn Cotton
Frank Ronan

The Girl at the Lion d'Or
Sebastian Faulks

An Emotional Optimist *(continued)*

◄ themselves and that they are very connected to other people. So you can be a complete loner and be happy, but the majority of people who are happy have a lot of links to others.

Also happy people give themselves credit for the good things that happen and blame the bad things on the outside world or on other people, which you don't do out loud of course because other people would hate you! But in their own minds that's what happy people do. You're still taking responsibility for things, you can still apologise to people but you consider something a one-off rather than deciding you're a bad person. You acknowledge a mistake without punishing yourself for ever about it.

Where did you start your research for the book and how long did it take?
I did it mostly one chapter at a time, not in the order that they are now, and each one required very different research. I realised about halfway through that this was nine different subjects, which could have been nine different books. Having studied psychology for years, I knew where to start and I did lots of work in the British Library, in University of London library and on my roof terrace. It was a really nice period because it took me back to being a student, that experience of being on the hunt for something, getting to one lot of references, then another, then another and suddenly finding the most brilliant thing that's exactly what you want and so satisfying. That makes me happy! The more difficult bit is deciding

what to include. It could have been twice the length and that's where it becomes more like journalism, deciding which bits are more interesting.

Was there anything really surprising in your research?
I like the jealousy and earlobes study because I wonder how they thought of it in the first place. I like the creativity of people's experiments. For example, measuring how long athletes smile for whilst getting their Olympic medals on the podium; how did they come up with that?

Would you ever write fiction?
I definitely want to write another non-fiction book but I know I couldn't write fiction. Like my seven-year-old self, I probably still respect novels more – I do think there's something really special about them though I read more non-fiction. It's that thing of admiring what you don't do: it can't be that hard to write a non-fiction book because I've done it but I admire fiction because it is mysterious to me. The hardest bit about writing the book was writing the stories; the science was easier. I found it a lot more challenging to get the sense of drama that I wanted than writing about the ins and outs of an experiment.

Who are the writers who have influenced you?
My father in terms of thinking it's possible to write a book, that it's a normal thing to do. Before I wrote my book Paul Martin was ▶

FAVOURITE
BOOKS ***continued***

Non-fiction

The Drowned and the Saved
Primo Levi

Love
Stendhal

How We Die
Sherwin B. Nuland

Counting Sheep
Paul Martin

The Art of Travel
Alain de Botton

Irrationality
Stuart Sutherland

The Essays: A Selection
Michel de Montaigne

The Expression of Emotion in Man and Animals
Charles Darwin

The Merck Manual of Medical Information

The newspapers (not strictly speaking books, but I still pile them up beside my bed until I've read them)

An Emotional Optimist *(continued)*

◀ my living literary hero and I wanted to write something as good as one of his. One of my favourite books is *Counting Sheep* partly because he's a big fan of sleep and so am I. Now I'm on lots of different panels with him and we're good friends. Of course he doesn't know he's my book hero; well, he does now! ■

A Day in the Life of Claudia Hammond

THE *TODAY* PROGRAMME wakes me at eight o'clock, but my freelance career is arranged around not having to get up early because I'm definitely not a morning person. I usually start my working day in my study, perhaps setting up interviews or writing a lecture. If it's sunny I sit on my roof terrace making notes, but the need for dead-heading can prove distracting. If I'm indoors I usually work listening to Radio 3 or to CDs of Daniel Barenboim.

If it's not raining I'll cycle to Broadcasting House to record or edit radio features. Having worked there for ten years, I do bump into a lot of people, so inevitably every visit involves quite a bit of chatting. But I like to convince myself that this counts as work because they're the same people who commission features from me.

My job involves a mixture of writing, broadcasting and lecturing so no day is typical, but in the afternoon I might be teaching American students who've come over to spend a term at Boston University's London base learning about the psychology of the big social issues in Britain.

If I'm feeling energetic I might go to the gym after work, but more likely I think of an excuse not to, such as the urgent need to go to Selfridges to look at clothes. Sometimes I've no choice but to do interviews in the evening, but more often I'll go out with friends. One regular monthly engagement is Crochet Club. It involves meeting up in various London bars to crochet, but mainly, I'll admit, to enjoy a bottle or two of wine. I ▶

'My freelance career is arranged around not having to get up early because I'm definitely not a morning person.'

A Day in the Life *(continued)*

◀ love the diversity of restaurants and theatres in London, but I also like nothing better than going home for supper, watching *ER* and having a bath. Much as I love sleep, I never manage to get to bed early and invariably the day ends with the midnight news on Radio 4. ■

The Joy of Penguins

by Claudia Hammond

THE DISTANCE IS the same as from the Shetland Islands to Nigeria, but starts with the driest place on earth and ends with toothpaste-blue icebergs and spectacular mountain peaks. Yet this is all in one country – Chile. My partner and I travelled the length of the country, combining several overnight bus journeys, the occasional train and a four-day ferry journey through the fjords. Six weeks later we reached our destination – Punta Arenas. Throughout the journey we visited fascinating places but we had a special reason for going to Punta Arenas – we wanted to see the penguins.

On the appointed day we found the right bus to take us out past a vast duty-free car shopping centre, suddenly appearing where you least expect it, as though Bluewater car park had been transported to the ends of the earth. The weather was cold and grey while we waited on the quay for one of the many boats which take tourists out to see colonies of Magellanic penguins. Strangely no one else was there – no tourists in search of penguins and no one selling tickets. Eventually a man wandered past and told us that for the first time in months the trips had been cancelled due to bad weather. We had to leave early next morning to cross the border to Argentina in time for a flight, so this had been our only chance to see them.

There is something special about penguins, something psychologists realised long before the documentary on emperor penguins, *March of the Penguins*, became a surprise Hollywood hit. When I was ▶

The Joy of Penguins *(continued)*

◀ researching laboratory experiments on joy I was struck by the number of studies which used videos of penguins in order to make people feel happy. Other methods of manipulating a person's emotions have been tried. Music would seem an obvious one, but the physiological changes associated with watching a film are in fact greater than with listening to music. There is also the problem of taste; not everyone likes the same music, so something that makes me feel ecstatic might not have the same effect on you.

With penguins, however, it seems that everyone is a fan. I've challenged several book festival audiences to confess to a dislike of penguins and no one has. In fact I don't even need to show them a video clip of penguins to raise a laugh – the simple mention of a film of penguins waddling and slip-sliding on the ice makes the whole audience smile.

Barbara Fredrickson has often used videos of penguins in her experiments on emotions. Just two minutes of penguin-viewing is enough to make people behave differently from when they have just watched a serene film of fields and streams, an unpleasant scene from the film *Witness*, a video of a climbing accident or that screen-saver where coloured sticks build up in a pile.

It is much harder than you might expect to find film clips which elicit the same emotion in every person, every time. Two American psychologists James Gross and Robert Levenson tested more than 250 films to discover which clips are the best at procuring each emotion. A clip from *Whose Line Is It Anyway?* topped the charts for amusement, but to be fair to penguins, they were not included in the test. The researchers found that their own intuitions about the best clips were often wrong. One of the few clips which made people laugh even if they had not seen the whole film was the fake orgasm in a café scene from *When Harry Met Sally*, but in general a person's reaction is stronger if they know the story of the film. Perhaps this is why penguins are so useful. You don't need a backstory. You don't need time to get to know the characters. They are just there.

In Fredrickson's studies the people who had seen the film of penguins completed a task by looking at the bigger picture instead of focusing on small details, suggesting that feelings of joy broaden a person's thinking. They were also asked to imagine themselves in a situation which elicited a particular emotion before compiling a list of statements of what they would like to do if they were experiencing that feeling. The people watching the penguin film were able to come up with far more suggestions than the others. This seems to be something to do with experiencing positive emotions; they make us look outwards, taking more factors into consideration. And the feeling they reported more often than in any of the other situations was the urge to play.

This led me to wonder why these strange creatures are so special for humans. Why was I prepared to travel thousands of miles to see them? And why do they make us feel playful? Clearly there's an absence of reasons to actively dislike penguins; they're not threatening, they don't have any of ▶

'Penguins seem almost childlike, a feature emphasised by their apparent vulnerability and their clumsy attempts at getting around, rather like toddlers learning to walk.'

The Joy of Penguins *(continued)*

◀ the characteristics which invoke feelings of disgust in humans. But what about the positive? On land we are attracted to their similarities to humans. Their upright nature is unusual for a creature of their size, matched by another wildlife-film favourite – the meerkat. Because penguins are so small they seem almost childlike, a feature emphasised by their apparent vulnerability and their clumsy attempts at getting around, rather like toddlers learning to walk. We can't help but smile, but it's an indulgent smile. They seem to have an inherently comic nature. Whether it's true or not, everyone loves the story of RAF pilots in the Falkland Islands reporting whole flocks of penguins toppling over backwards because they were so fascinated by planes flying overhead.

In the most memorable penguin film I've seen, the penguins stood in a group waiting to dive into icy water, knowing that water contained danger. They would gradually edge nearer to the front and nudge each other on. In such a human way it seemed that no one wanted to be first to take the plunge. In Stanley Park in Vancouver the penguins have a series of diving boards. It's common to see a penguin climb all the way to the top of the highest board, look over the edge, think better of it and try a lower board instead. Each board is tried in turn and rejected, until finally the bird walks round to the side of the pool and simply jumps in.

There is another human element we find attractive in penguins – they are monogamous and take it in turns to look after the egg. It's the very fact that these responses to penguins seem to affect us all which has made them so useful to the psychologists with the task of finding out exactly how joy affects the mind and the body.

So there I was at the other end of the world with no penguins to see. We tried to book ourselves onto a trip by mini-bus to see another colony, but the tourist office told us they were cancelled too. Punta Arenas may be hard to get to, but it has some great clothes shops, so first we had a little retail therapy followed by another activity which makes us happy, but this one might not be universal, and that's air hockey. After a few vicious games we returned to our guesthouse and told the owner what had happened. Defeated, we went back to our room. A knock on the door. The owner warned us to wear everything we had. A mini-bus would pick us up in three minutes and it was going to be cold.

The area where the small Magellanic penguins had their nests was separated from the beach by a path which took us to wooden viewing platforms. It seemed as though there were only a few pairs until we looked down from the platform and realised there was movement everywhere. Hundreds of penguins with smart black and white striped faces were huddled in pairs, each with a fluffy baby between their feet. Others lined up and waited for us to pass before they would cross the path on their way out to sea, like people waiting at a pelican crossing. Then of course they would watch the water and wait, waiting and waiting until someone else launched themselves into the water first. We were freezing, but we felt just one emotion – joy. ■

> ‘Everyone loves the story of RAF pilots in the Falkland Islands reporting whole flocks of penguins toppling over backwards because they were so fascinated by planes flying overhead.’

If You Loved This, You Might Like . . .

Mutants
Armand Marie Leroi
Armand Marie Leroi's witty and clever book details the development of the human body and how humans are created. His research dovetails neatly with Hammond's, especially on the nature of beauty, love and disgust.

The Sad Truth About Happiness
Anne Giardini
A novel in which the main protagonist, Maggie, begins a search for true happiness after she learns that unless she finds it, she will die in three months. An interesting study of what it is that truly makes us happy.

Authentic Happiness
Martin Seligman
Famous for his book *Learned Optimism*, Seligman continues his focus on positive psychology in this much praised book about what true happiness is. Accompanied by a thorough and informative website and grounded in years of academic study, the book provides a fascinating and readable approach to the most elusive and desirable of feelings.

Emotion: the Science of Sentiment
Dylan Evans
A research philosopher at King's College London, Evans examines some of the current thinking about the science of emotions and discusses how emotions can help us function better as humans, as long as we pay them due attention.

The Man Who Mistook His Wife for a Hat
Oliver Sacks
The famous neurologist's most celebrated book, in which he details various case histories which are illuminating in themselves but also in terms of human experience. Sacks, who has been called the 'poet laureate' of medicine, draws the reader into the particular lives of those affected by neurological disorders and enables us to see their humanity as well as our own.

Emotional Intelligence: Why It Can Matter More than IQ
Daniel Goleman
Our success and future used to be determined by our IQ: the higher it was, the better our chances. But in recent years the importance of our emotions and our EQ has been under discussion: the way we use and learn from our emotions is just as significant for our ability to do well at work or in relationships. What's more, unlike IQ, EQ is not fixed and can be improved.

Making Happy People
Paul Martin
Like Claudia Hammond, Paul Martin uses a scientific background to explain the most desirable feeling of all: happiness – what it is, why it matters, where it comes from and how to raise children to achieve it, as well as find it as adults. ▶

FIND OUT MORE

Surf

www.authentichappiness.org
The website of Dr Martin Seligman, an eminent psychologist with a particular interest in positive psychology. Resources include quizzes about happiness, information and suggestions on how to be more hopeful, joyful and less fearful.

www.quackwatch.org
The website which gives you the real story about any unusual medical treatment or therapy you might be considering. See under 'mental help' for questionable therapies for the mind.

http://human-nature.com
A site run by a US academic, which comes mostly from an evolutionary perspective. It has all sorts of articles on the mind as well as reviews of books, and TV and radio programmes on the subject.

If You Loved This *(continued)*

◀ ***Happiness***
Will Ferguson
What would happen if a self-help book actually worked, if the route to happiness could be found between the pages of a paperback and everybody who read it became rich, loved and vice-free? When Edwin de Valu publishes such a book, little does he know how ungrateful the world will be.

Lightning Source UK Ltd.
Milton Keynes UK
UKHW041314231118
332829UK00007B/586/P

9 780007 164677